"Business travel is like  ou
never know what's g   is
– sometimes you hav   p
outside your comfort z   it
her escapades, taking   d
charming way."

Sumita Shah, **Management Consultant**

"Travel may broaden the mind, and it certainly oils the wheels of global commerce, but doing it for a living is not for everyone. Gillian's book is a lively personal account of the challenges as well as the rewards of business travel, drawn from her extensive experience as a solo female traveller."

John Davies, **Company Director**

"Who would have thought that the exploits of an accountant abroad would make for such a great read. At breakneck speed, the narrative whisks our hero, Gillian, from continent to continent as she makes the best of the unexpected (sometimes scary, often funny, always dramatic) situations that come her way. Gillian relates her adventures with flair and brevity, often leaving me laughing out loud at the craziness of it all."

Penny Steed, **Retired NHS Manager**

"In 'Unconventional Routes: Around the World from Boardrooms to Backpacks' Gillian takes us on an extraordinary voyage that transcends the confines of traditional business travel, revealing the untold adventures of a female accountant with an insatiable appetite for the world. With a blend of wit, charm, and raw honesty, Gillian navigates through the unexpected twists and turns of visiting some of the most challenging destinations on the planet. Her stories are more than mere travel logs; they are vivid illustrations of resilience, adaptability, and the sheer joy of embracing the unknown."

Mitzi Wyman, **Director**

UNCONVENTIONAL Routes

Around the World from
BOARDROOMS to **BACKPACKS**

GILLIAN FAWCETT

The Book Guild Ltd

First published in Great Britain in 2024 by
The Book Guild Ltd
Unit E2 Airfield Business Park,
Harrison Road, Market Harborough,
Leicestershire. LE16 7UL
Tel: 0116 2792299
www.bookguild.co.uk
Email: info@bookguild.co.uk
X: @bookguild

Copyright © 2024 Gillian Fawcett

The right of Gillian Fawcett to be identified as the author of this
work has been asserted by them in accordance with the
Copyright, Design and Patents Act 1988.

All rights reserved. No part of this publication may be
reproduced, transmitted, or stored in a retrieval system, in any form or by any means,
without permission in writing from the publisher, nor be otherwise circulated in
any form of binding or cover other than that in which it is published and without
a similar condition being imposed on the subsequent purchaser.

Typeset in 11pt Minion Pro

Printed and bound by CPI Group (UK) Ltd, Croydon, CR0 4YY

ISBN 978 1835740 354

British Library Cataloguing in Publication Data.
A catalogue record for this book is available from the British Library.

To Aidan who inspired me to write this book.
This is for you.

CONTENTS

Abstract   xi
Introduction   xiii

1. A Stranger from Hanoi Saves the Day   1
2. Swimming with Sharks in Belize   10
3. OMG – Out of Africa   15
4. *Dobry Den,* Moscow   19
5. Kenya Rhino Baby   24
6. Who Would Say No to the Caribbean?   29
7. South Korea – Kimchi and the Kitchen Battle   35
8. A World Cup Qualifier in Buenos Aires!   40
9. Bangladeshi Fever   50
10. Sri Lanka Sea Breeze   58
11. Tuk-Tuk Thailand   62
12. Bedbug China   70
13. Maldives – Wigs Away!   76
14. The Brussels Hops   83
15. Kathmandu – My Not-So Shangri-La Experience   86

16.	Awesome Canada	93
17.	Springbok in Johannesburg	98
18.	Xi'an, China: Deer Goulash and Disappearing Delegates	102
19.	Dressed to Impress for the Bengal Club, Kolkata, India	115
20.	Thriller in Manila	119
21.	Barbados, Here I Come!	122
22.	Jamaica – Coffee Beans and Independence	125
23.	Overdressed, Washington, DC	128
24.	Running up the Spanish Steps	133
25.	Northern Ireland – Careful What You Wear	136
26.	Heading for the United Nations, Geneva, Switzerland	139
27.	Tokyo Revisited	142
28.	Return to Kenya – The Bodyguard and Me	147
29.	Goulash in Budapest	153
30.	Panic in Islamabad	157
31.	Malaysia – Set Those Birds Free	163
32.	Living it up in Paris	168
33.	Hot in Hyderabad	170
34.	Small Pleasures in Hong Kong	175
35.	My Old Friend, Dhaka	181
36.	Thriller in Manila, Continued	184
37.	Punching the Air in Zimbabwe	191
38.	A High Red Blood Cell Count in Nepal	196
39.	Cinnamon in Sri Lanka	209

40.	One Last Time in Beijing	212
41.	Full Circle to Hanoi	221
42.	Reflections in the COVID Years	227
43.	Lost Luggage in Bosnia and Herzegovina	229
44.	The Pristina Bears of Kosovo	235
45.	Travelling in My Shoes	240

| Acknowledgements | 243 |
| My Business Travel Schedule 2009–2023 | 245 |

ABSTRACT

Being a businesswoman on a whirlwind tour of forty-three countries over the span of a decade is no ordinary adventure. It's a wild ride through the ups and downs of global exploration. Let me tell you, it's been a rollercoaster of experiences that have left me wiser, more adaptable, and occasionally questioning my sanity.

Understanding and embracing differences became my unofficial language during these travels. I've learnt to navigate the intricate tapestry of cultures and integrate myself into the vibrant fabric of foreign lands, even if it meant stepping out of my comfort zone and facing awkward situations. Trust me, there were moments when I felt like a square peg in a round hole, but hey, that's the price you pay for expanding your horizons.

No two journeys were ever alike. Each had its own unique flavour, spiced with a generous sprinkle of incidents that ranged from side-splitting hilarity to spine-tingling unease and, on occasion, hair-raising danger. It was like playing a never-ending game of travel roulette, never knowing what kind of adventure awaited me at the next destination.

Picture this: laughing uncontrollably with a group of colleagues over a comically lost-in-translation moment, only to find myself nervously glancing over my shoulder in a sketchy neighbourhood the very next day. It was a constant dance between laughter and unease, keeping me on my toes and ensuring that boredom never had a chance to set in.

Buckle up and get ready for a whirlwind journey through my escapades. Together, we'll laugh, cringe, and gasp at the absurdity and beauty of the world I've had the privilege to explore.

INTRODUCTION

I have often been hounded by my family and friends, insisting that I write a light-hearted journal about my globetrotting adventures as a businesswoman. It seems there are countless books penned by men chronicling their journeys, but there's a distinct lack of women sharing their escapades. So, after much contemplation, I've decided to put pen to paper. Though, let's be honest, with all the distractions around, it's harder to start than resisting the urge to eat that last slice of pizza.

Most people assume that a life of constant jet-setting is all glitz and glamour. But let me tell you, it's not all champagne sipping in plush business-class seats while ticking off dream destinations from your 'bucket list'. Oh no! It's hard work, even with a never-ending flow of bubbly and visits to countries that make others green with envy.

I was in my early forties when I joined a global professional accountancy organisation, after an unconventional career from parliament to regulation. Back then, I thought I had struck gold – a dream job that combined policy, finance, and trotting across the globe. Plus, I had the freedom to come up with ingenious ideas for promoting and developing the accounting profession in the public sector. It was like being handed a golden ticket to soar into the skies of opportunity. I was ready to take off like a rocket!

Little did I know that my life would soon be filled with long-haul flights that messed with my body clock, racing through airports at breakneck speed to catch connecting flights, and grappling with the woes of jet lag. Forget about a routine, balanced diet, and a good night's sleep. And let's not even get started on the increased odds of developing deep vein thrombosis. Prior to every trip, I would double and triple-check my visas and vaccinations. By the time I stopped travelling for work, my vaccination card looked like a comprehensive encyclopaedia of diseases: yellow fever, tetanus, typhoid, hepatitis B, diphtheria, and even malaria! Phew!

Each journey required military-grade planning. I would frantically pack all the business essentials: laptop, charger, business cards, snazzy attire, toothbrush, and, of course, a spare pair of knickers. There was one time I even resorted to googling an internet checklist just to ensure I didn't forget anything vital. Guess what? The list had a whopping thirty items! I eventually gave up and decided that a toothbrush and a backup pair of undies would suffice. After all, I could always buy the rest when the plane landed.

As a female traveller, I understand the nagging feeling of being less safe than our male counterparts on business trips. But you know what? Some of my hair-raising moments were a result of my own poor planning – like booking flights that arrived in the dead of night. I probably should have taken better care of myself. While on the road, I encountered my fair share of gender discrimination and inequality, which made adjusting to different cultures and customs a bumpy ride. There were moments when I had to clench my teeth and swallow my frustrations.

That being said, throughout the years, I've had the time of my life in countless countries. I consider myself incredibly fortunate to have travelled extensively. Business travel has given me a unique chance to immerse myself in diverse cultures, understand people's values and beliefs, and gain insights into the intricate tapestry of global economies and politics. It's like being a tourist on steroids, absorbing more knowledge than a quick hop-on, hop-off visitor ever could.

Within the pages of this journal, you'll find tales ranging from the side-splittingly funny to the spine-tinglingly scary, and everything in between. I'll introduce you to some of the most intriguing characters I've encountered along the way – people who will make you laugh, scratch your head, and question the very fabric of reality.

This journal is dedicated to three special entities in my life: my dear mum, Mary; my loving husband, Aidan; and Officer Dibble, our patient feline friend who faithfully awaited my return from every adventure. It's also a tribute to those saintly and eagle-eyed friends who tirelessly reviewed my drafts, probably contemplating pulling out their own hair due to my atrocious grammar and never-ending typos.

1. A STRANGER FROM HANOI SAVES THE DAY

**London Heathrow Airport –
Suvarnabhumi Airport, Bangkok, Thailand –
Nội Bài International Airport, Hanoi, Vietnam**

Flight time: Sixteen hours

31 May 2009

Picture this: Suvarnabhumi Airport in Bangkok, the stage for my very own sprinting Olympics. I'm dashing through the terminal like Dina Asher-Smith on a caffeine-fuelled rampage, desperately trying to catch my connecting flight to Hanoi for an international accountancy conference. My heart is pounding in my chest, and my feet feel like they're tap-dancing on hot coals.

With my overstuffed briefcase threatening to burst at the seams, I hurdle over unsuspecting travellers, zigzag through crowds, and dodge a rogue luggage cart with the agility of a ninja. The airport concourse seems to stretch out like an endless marathon track, testing my stamina and determination.

Tick-tock, the clock is ticking mercilessly. I glance at my watch, sweat dripping from my forehead, as I realise

I have mere minutes left to make it on board. Every step feels like an eternity as I race against time.

And then, like a scene from an action movie, I spot the gate in the distance. It's like a beacon of hope, urging me to keep pushing forward. With sheer willpower, I muster up the last ounce of energy in my exhausted body and sprint towards the finish line – the cabin door.

As I reach the gate, gasping for air, I see the flight attendants preparing to close the door. Adrenaline surges through my veins, granting me superhuman speed as I make a desperate lunge. Just as the door threatens to seal my fate, I slip through like a contortionist escaping a straitjacket.

Collapse. I collapse into the nearest seat, my chest heaving, my briefcase a crumpled mess at my side. I'm a victorious warrior, triumphant in my battle against time. Sweat-soaked and panting, I take a moment to catch my breath, feeling the rush of accomplishment coursing through my veins.

1 June 2009

Good morning, Vietnam! Or at least, that's what I'm trying to convince myself as I stumble out of bed, nursing a sore throat caused by the Arctic-level air conditioning in this opulent penthouse suite. I can't shake the feeling that the receptionist made a mistake and put me in the wrong room. All evening, I've been plagued by visions of being unceremoniously evicted by an irate guest. As if that weren't enough, my period has chosen this very moment to make an appearance, and, of course, I forgot to pack any tampons.

With a mere thirty minutes until the business meeting, I find myself in a race against time to track down those elusive tampons. Stepping out onto the chaotic streets of Hanoi feels like entering a real-life version of Frogger, as I desperately dodge the onslaught of screeching motor scooters. With an estimated 5 million bikes and scooters swarming through the city, it's a symphony of traffic chaos and a nerve-racking experience. Surviving the crossing becomes a personal achievement. The humidity hangs heavy in the air, mingling with the suffocating fumes and the cacophony of honking horns. But I soldier on, determined to find a kiosk that can save me from this period-induced predicament.

In most places I've travelled, I've been able to find someone who speaks at least a bit of English. But not this time. The shop assistant behind the counter seems utterly uninterested in deciphering my desperate gestures and my Vietnamese vocabulary is limited to 'hello' and 'thank you'. She gives me a perplexed look and proceeds to place every single item in the kiosk on the counter, as if hoping I'll miraculously find what I need amongst the random assortment of products. With time ticking away, I'm left with no choice but to resort to my artistic skills – or lack thereof. I grab a piece of paper and draw what can only be described as a wonky, rocket-shaped object. She smirks, nudges her colleague, and they burst into laughter. Well, at least my drawing skills can provide some entertainment. Thankfully, the dong drops, and she discreetly retrieves the much-needed merchandise from beneath the counter, securely hidden in a plain brown paper bag.

Now armed with the necessary supplies, I rush to the meeting that's about to start in a stuffy, overcrowded office in downtown Hanoi. The humidity sticks to my suit like a clingy ex, and the incessant whirring of an electric fan above my head threatens to turn my stomach. My fingers feel sticky as I desperately cling to my notes, and the room is filled with sweaty bodies, some of whom look half-asleep in the sweltering air. Pearls of sweat trace a slow path down my spine, collecting at my lower back before making their grand escape. I struggle to make sense of the interpreter's questions, which seem to revolve around the groundbreaking idea of having a finance director who is professionally qualified within a company. It's been a long, long morning, and I can't help but question why on Earth I've travelled halfway across the world to state the obvious. How did my career come to this, you ask? Well, that's a question I'm still trying to answer myself.

2 June 2009

Jet lag has definitely knocked me off my game this morning. After a gruelling sixteen-hour flight and a pesky six-hour time difference, my body clock is screaming at me to crawl back into bed. Yet here I am, expected to work and function like a normal human being. And to add to the fun, my local office colleagues are a bunch of energetic workaholics. Vietnam seems to have an inexhaustible supply of young, energetic folk, with a whopping 60% of the population under thirty. These colleagues of mine start their day at the ungodly hour of 7am and continue burning the midnight oil. Naturally, this means that I must be available to them twenty-four-seven, because who needs sleep, right?

Today's adventure involves sharing a podium with speakers from Russia and Cambodia. As I glance around, it dawns on me that I'm not only the sole presenter from Western Europe, but also the only female on the panel. Now, how does that fit into their special political and diplomatic relations? Your guess is as good as mine. I finish my presentation and turn to the chair, who is shamelessly snoozing away. With a gentle nudge, he jolts awake, sending his glasses tumbling to the floor, much to the delight of the audience who erupt in applause. Classic!

Vietnam News covers the conference highlights, but guess whose contribution gets excluded from the article? Yours truly. Just when I thought I had gracefully made it out unscathed, I suddenly find myself swarmed by a horde of journalists and TV crews. Oh joy, the fun never ends.

I'm hastily guided to a sofa and instructed to take a seat under a massive potted palm for a live TV interview. And then, out of the blue, comes the million-dollar question: "Well, Ms. Fawcett, what do you think the Vietnamese government should do about the economy?"

Surprise! My face turns a delightful shade of beetroot, and I resemble a scared rabbit caught in the blinding glare of headlights. Seriously, answering such a question is way above my pay grade. Last time I checked, I wasn't the Chancellor of the Exchequer. So, with my heart pounding, I gather my wits and take a deep breath. "Umm, you know, uh, maybe invest in education? Yes, education! That's what you need for a strong economy." Miraculously, it seems like the journalists are nodding their heads in agreement or at least acknowledgment. Phew, dodged a bullet there.

Ah, the joys of being thrust into unexpected situations and having to think on your feet. Who needs a script anyway? I'm just here to add a touch of chaos to the proceedings. And so, my adventures continue, filled with jet lag, surreal moments, and the occasional spotlight. Stay tuned for more thrilling episodes of *The Misadventures of Ms. Fawcett, the Accidental Economist*.

3 June 2009

Oh my goodness, getting out of bed this morning feels like an impossible task thanks to this relentless jet lag. And to top it off, I can't help but blame my late-night indulgence of watching CNN in bed at 2am, accompanied by a Cadbury's Dairy Fruit and Nut chocolate bar. Note to self: midnight chocolate cravings and news binges do not mix well with combating jet lag.

As I peruse this morning's newspapers, I discover an interesting mix of headlines. The European Union's commitment of a whopping $994 million in aid to combat environmental degradation in Vietnam takes the spotlight, as it should. However, I can't help but chuckle at the unexpected juxtaposition of news about the UK MPs' expenses scandal. The journalist seems to have gone into meticulous detail about everything from reimbursing expenses for horse manure to tennis court repairs. Talk about local news taking a strange turn. Don't they have more pressing matters to report on? Oh well, the world works in mysterious ways.

And here we are, on the penultimate day of this whirlwind adventure, with me speaking at a conference on financial accounting standards. Let me just say, it's not exactly my area of expertise. So, you can imagine my panic when I discover that the local office has agreed to have me present on this mind-numbingly dry and technical subject for a whole day. Lost in translation, much? There's no way on this beautiful planet that I can keep an audience engaged for seven excruciating hours with such riveting content.

But lo and behold, a stranger named Ken comes to my rescue like a knight in shining armour. This man is a true gem, a walking encyclopaedia of financial accounting knowledge. Just by the sight of him clutching a thousand-page financial accounting manual under his arm, I know I'm saved. Ken is an absolute wizard when it comes to diving into the nitty-gritty details, and he pulls me back from the brink of a meltdown. Together, we run the session like a seamless tennis rally, effortlessly returning each other's serves (or rather, answering questions) with great finesse.

A heartfelt thank you to, Ken, my saviour in Hanoi. Without him, I might have been lost in a sea of technical jargon, drowning in my own lack of accounting expertise. It's in moments like these that strangers become allies, and unexpected heroes emerge to save the day. Here's to you, Ken, the unsung hero of financial accounting standards.

4 June 2009

Hip, hip, hooray! Today is finally 'me' time, a chance to explore and indulge in the wonders of Hanoi.

As the capital of Vietnam, Hanoi has risen like a phoenix from the ashes of the Vietnam War, now a vibrant city brimming with resilience and pride. Its streets are adorned with large, tree-lined boulevards and impressive architecture that bear witness to its periods of French and Chinese occupation. However, it's not the grandeur of the buildings that first catches my attention, but the symphony of noisy scooters that dominate the roads, swerving and honking as they navigate their way through the bustling streets. Five days into my stay, and I still haven't quite mustered the courage to confidently cross these chaotic roads.

In search of tranquillity, I escape the maddening traffic and find myself at the shores of Hoan Kiem Lake, a picturesque emerald-green lake nestled in the heart of the city. Walking along its perimeter, I'm captivated by the extraordinary hue of the water, a result of the sunlight filtering through the lush greenery that surrounds the lake. It's a serene oasis where people engage in gentle exercise, savour cups of tea, or enjoy a refreshing local beer. And just beyond the lake, I stumble upon a women's museum that piques my interest.

Inside, a vibrant exhibition of old propaganda posters from the 1970s catches my eye. The posters depict women engaged in agricultural activities, offering a fascinating glimpse into the past. Oh, how I wish I could discreetly sneak one of those captivating artworks back home with me!

In a smaller gallery within the museum, I discover an equally intriguing display that sheds light on the plight of street vendors in Hanoi. Most of these vendors come from impoverished rural areas, where farming income alone is insufficient to sustain their families. They flock to the city to sell their produce, desperate to make ends meet. The information highlights government proposals to restrict their presence in the city centre, viewing street vending as a backward practice. The clash between planners, politicians, and ordinary people over the use of contested urban space becomes evident. However, the Women's Museum has taken a stand, organising a petition in support of the rights of street vendors. Naturally, I'm eager to add my signature as a show of solidarity.

Returning to Vietnam a few years later, I would learn that despite the petition, the legislation passed. However, due to the street vendors' defiance and determination, they managed to reclaim their place in the city.

Bravo to their tenacity and courage, standing tall against the odds and proving that their presence is an integral part of Hanoi's vibrant tapestry.

Cheers to the street vendors, the unsung heroes of the city streets, who add a dash of colour, culture, and livelihood to the urban landscape. Their spirit and resilience are an inspiration to us all.

2 SWIMMING WITH SHARKS IN BELIZE

London Heathrow Airport – Miami International Airport, US – Philip S.W. Goldson International Airport, Belize

Flight time: Twelve hours

18 February 2010

It's 5am, and my colleague Monica and I are zooming through the ramshackle streets and run-down neighbourhoods of Belize City in a sleek, black limousine. We're on our way to Love FM for a live TV and radio interview – quite the early morning adventure if you ask me.

As we speed along, I can't help but wonder what possessed Monica from the Caribbean office to agree to this interview. I mean, who in Belize City would want to listen to accountants discussing financial reporting while munching on their breakfast? Are people even awake at this ungodly hour? I'm still rubbing the sleep out of my eyes as we glide through the streets.

We arrive at the studio, greeted by the lively tune of Aretha Franklin's song, 'Who's Zoomin' Who?', and a presenter who's channelling his inner dad dancer. He shoots us a playful wink before making his way over to introduce himself. With smiles on our faces, we mentally prepare ourselves for whatever is about to unfold.

After the obligatory greetings and patronising introductions as 'ladies from the UK and Trinidad', the presenter enquiries about our presence in Belize. Monica promptly fills him in on the international accounting conference we're attending. Then comes the zinger: "Gillian, what are your thoughts on the Belize government's failure to publish a set of accounts in the last ten years?"

In a diplomatic dodge, I improvise by stating that, "Every government, regardless of location, should prioritise financial transparency." It may not have directly addressed his question, but hey, it'll do.

The host moves on to a broader topic, asking, "Gillian, Monica, do you believe women accountants make better bean counters than men?" We exchange puzzled glances, unsure where he's going with this line of questioning. However, Monica manages to deliver a solid response. Now fully awake, we anticipate a short break, though we're taken aback by the choice of music: Donna Summer's 'I Feel Love'. It's mind-boggling how this 1980s disco tune fits into the interview. Either the listeners are in a comatose state or dancing up a storm by now.

Resuming the interview, the host throws another curveball our way, asking, "Ladies, do you think accountants are boring bean counters who avoid taking risks?"

Ah, my chance to shine on national radio and TV! Without missing a beat, I blurt out, "Nonsense, just yesterday, I was swimming with sharks off the shores of

Caye Caulker!" Now, this tiny island has a baggy shirt, bear feet and laid-back mentality and is surrounded by a barrier reef teeming with sea life, including sharks.

Little did I know that my confident statement would cause the host to burst into hysterical laughter, broadcasting to the entire population of Belize, "Yes, but these are nurse sharks with no teeth, though they can give you one hell of a suck." Oh dear, I've been rumbled on Love FM in Belize.

Drowning in embarrassment, we're swiftly chauffeured back to the hotel as the sun rises over the rooftops, promising yet another scorching day. As we pull up, we're met with a scene of police officers surrounding the premises. Apparently, a resident had been shot dead in a drug-related incident. It's now 7:30am, and all I want to do is crawl back into bed, praying that no one from the conference caught wind of the interview. Unfortunately, fate had other plans, and the interview, complete with disco tunes, plays on an endless loop throughout the day, providing endless amusement to our conference colleagues.

19 February 2010

Today kicks off with a delightful visit to the central bank, where I'm tasked with running a seminar on governance and ethics for its esteemed staff. Now, let me tell you, this is no ordinary building – it's an imposing structure that houses an astonishing collection of art. Talk about a workplace with a touch of class. As we begin with a tour of the bank's art spaces, I find myself marvelling at a plethora of colourful oil paintings, whimsical drawings, captivating prints, and photographs. Forget the ethics seminar for a moment; I'm just here to soak in the artistic wonders.

Now, it's not uncommon to find art in central banks, but not all banks take it to the level of the Belize Central Bank. This place has turned art collection into an art form itself. You'd think they have a dedicated team of investment gurus managing their artistic treasures, but guess what? According to our guide, they don't! They neither sell the art nor have a fancy investment strategy for it. They simply embrace art as a means of connecting with the society they serve as a public institution. And hey, it's not just for the stuffy bankers; the art is here for everyone's enjoyment, staff and visitors alike. Kudos to the Belize Central Bank for putting creativity and culture at the forefront.

Now, hosting art collections in central banks isn't always smooth sailing. They've faced their fair share of controversies and scandals. Remember that time they had to remove nude paintings due to staff complaints? Well, fear not, I haven't stumbled upon any scandalous nudes during my art-filled escapade here.

After a fantastic start to the day, we finally dive into the meat of the matter – the governance and ethics workshop. We tackle some juicy ethical dilemmas that bank staff may encounter, from the treacherous realm of insider trading to the delicate art of managing conflicts of interest. It's a lively discussion, full of timely insights, especially since the bank is gearing up to review its ethical code. Oh, the drama! But hey, it's all in the name of creating a more responsible and ethically sound financial institution.

So there you have it, a day filled with art, ethics, and a sprinkle of controversy. Who said banking couldn't be exciting? It's all about balancing the books and enjoying a splash of creativity along the way.

3. OMG – OUT OF AFRICA

Philip S.W. Goldson International Airport, Belize –
Miami International Airport, US – London –
O.R. Tambo International Airport, Johannesburg, South Africa –
Harry Mwanga Nkumbula Airport, Livingstone, Zambia

Total flight time: Thirty-one hours

22 February 2010

As the beautiful orange sun begins its descent over the Zambezi River, I find myself in pure bliss, savouring a mouth-watering 'Out of Africa' cocktail – a delightful concoction of grapefruit, grenadine, oranges, pineapple, safari liqueur, and vodka. Life couldn't get any better than this.

Drifting lazily down the expansive river, our boat becomes a wildlife-seeking vessel, scanning the banks for lurking crocodiles, bathing hippos, and a colourful array of birdlife. It's like heaven on water. Accompanying me on this adventure are my colleagues from various countries and organisations, and let me tell you, we're all in absolute awe of this majestic, sluggish river. At some points, it stretches up to a whopping two kilometres wide! Just imagine that.

The Zambezi River originates a staggering 1,200 kilometres upstream in the northwestern Province of Zambia, and as it races through a series of dramatic gorges it eventually finds its way to Lake Kariba before continuing its journey all the way to the Indian Ocean. Ah, the wonders of nature. It's moments like these that make me forget all about the traumatic visa fiasco I faced upon arrival.

Let me recount the tale of my arrival: after enduring a gruelling thirty-one-hour flight from Belize, with a quick pit stop in London, I arrived in Livingstone, Zambia, without a visa and any single means to pay the $50 entry fee. Exhausted and barely conscious, I was plucked from the border control queue and whisked away to the nearest cash point. Alas, fate had other plans, as the cash point decided to take a lunch break just when I needed it most. With no other options in sight, it seemed like my only option was to be put on the next flight back home. But lo and behold, the Zambezi government came to my rescue, kindly paying the entry fee. I suspect it had something to do with the realisation that without this small act of goodwill, they would lose one of their precious few female international speakers for their conference. With my visa finally stamped, I was herded along with the other delegates, accompanied by a convoy of minibuses and a chorus of police sirens. Now that's what I call an entrance.

Leaving behind Livingstone, a town that owes its existence to the arrival of a railway line in the early twentieth century, we make our way towards the enchanting Victoria Falls. To my absolute delight, we arrive at a luxurious resort situated next to the falls themselves – the famous 'smoke that thunders'. Life couldn't be sweeter.

However, my heart goes out to the unsuspecting holidaymakers who probably spent a small fortune for their trip of a lifetime, only to find themselves sharing their piece of paradise with hundreds of accountants swarming around the swimming pool. They certainly didn't expect the resort to be taken over by delegates proudly sporting T-shirts proclaiming 'Now is the Time for Internal Audit'. Neither did I, to be honest. I can only imagine the confusion and disbelief as they flipped through their holiday brochures, wondering what happened to their tranquil getaway.

But fear not, for the holidaymakers had their revenge. Oh, how they laughed as they witnessed me single-handedly fending off a ferocious baboon, its dirty, yellow teeth gnashing as it tried to snatch my laptop. I emerged victorious, while the baboon scurried away, clutching a sugar bowl instead. The holidaymakers captured the action, undoubtedly grateful for the unexpected entertainment. Baboons soon became a constant nuisance, a reminder for me to always close windows and doors, lest these mischievous thieves will ransack my room in search of goodies.

Now, the hotel where I find myself is just a stone's throw away from the majestic Victoria Falls. And let me tell you, when I first lay eyes on them, words fail me. They are an absolute spectacle, living up to their reputation as one of the seven natural wonders of the world. The falls mark the division between the upper and lower Zambezi as the water cascades a staggering one hundred metres into a deep, narrow gorge. It's the widest curtain of water in the world. Spray soars into the air, enveloping everything in a fine mist. As I emerge from the haze, soaked to the bone, I pose for a photograph next to the Cecil Rhodes statue. Looking back, I realise it may not have been the wisest choice, considering his notorious reputation as an imperialist and racist. It would be years later before statues of his likeness were removed from both the UK and South Africa. Ah, the twists and turns of history.

But let's return to our idyllic cruise along the Zambezi River. Stepping off the boat onto a small island in the middle of the river, clutching my second sublime 'Out of Africa' cocktail, I encounter an unexpected sight. An elderly woman, resembling an aging film star with her headscarf and vibrant crimson lipstick, suddenly erupts into a panic. "Get off the beach!" she screams. Now, this same woman had been quite pleasant earlier, sharing stories of her upbringing in former Rhodesia. So, I'm a tad confused as to what has rattled her cage so fiercely. She appears extremely alarmed, her face contorted with fear as she continues shouting. Reacting quickly, I drop my precious cocktail and rush to the centre of the island. As it turns out, she's watching out for my safety, as those pesky Nile crocodiles can crawl out of the river and snatch unsuspecting beachgoers – probably those who've had a few too many cocktails.

Oh, the adventures and unexpected encounters I have faced on this trip. From visa dramas to baboon battles, from the grandeur of Victoria Falls to the peculiar warnings of an island guardian. This is the stuff memories are made of.

4. DOBRY DEN, MOSCOW

**London Heathrow Airport –
Sheremetyevo International Airport, Moscow, Russia**

Flight time: Four hours

20 May 2010

The moment the plane lands at Sheremetyevo International Airport in Moscow, I'm greeted by a chauffeur and swiftly whisked away in a luxurious limousine, zooming through the city streets at breakneck speed. This unexpected downtime provides me with a precious opportunity to gather my thoughts before the upcoming international public finance conference. I ponder over my speech, contemplating whether I should delicately touch upon the struggling Russian economy and the widespread poverty that plagues the nation. The last thing I want is to ruffle any feathers and upset our hosts.

Now, some folks still perceive Russia as a communist country, but that hasn't been the case since 1991 when the Soviet Union collapsed. Led by none other than Vladimir Putin, it has transformed into a kleptocracy – a system perpetuating itself and catering to those at the pinnacle of power who utilise corruption as a means of domestic control and projecting influence abroad. Transparency International reveals that corrupt officials and politically connected individuals have been laundering astronomical sums of money and stashing their ill-gotten gains overseas for more than two decades. And here's a lesser-known fact: Russia, with its grand ambitions of being a military superpower, still has one-fifth of its population lacking access to indoor toilets. Quite intriguing, isn't it? But perhaps it's best to steer clear of such details in my speech. I wouldn't want to sour the conference.

As we glide through the city streets, I cast my gaze out of the car window and take in the architecture. Oppressive, brutal, and predominantly grey – these are the words that come to mind. Despite it being early summer, the streets exude a dingy and unwelcoming atmosphere. Some people may wax poetic about the city's majesty, but I confess that I fail to grasp its allure, even as we pass by the iconic Red Square and the formidable Kremlin. However, luck seems to be on my side as I discover that I will be staying in one of the renowned 'wedding-cake' buildings, now transformed into a Hilton Hotel. These magnificent structures, constructed in the 1950s to Stalin's exact specifications, resemble tiered wedding cakes, an attempt to modernise and reshape the city's skyline in the aftermath of World War II. Upon arrival at the hotel, I'm greeted by an opulent foyer adorned with lavish marble and a grand staircase. However, it becomes apparent that the dark brown furniture and worn carpets have seen better days. Ah, the passage of time.

Despite the mixed impressions and the weighty considerations for my speech, I remain optimistic that this conference will provide valuable insights and opportunities. After all, there's always something fascinating to uncover in every corner of the world, even in the most unexpected places.

21 May 2010

This morning I have the pleasure of meeting Vladimir, the head of the Moscow office, for the first time. Together, we make our way to Moscow University for the international conference. To my surprise, upon arrival, I discover that I am the sole international and female speaker. It seems a pattern is emerging – I may just be a token speaker on both counts.

We are ushered into a small, smoke-filled room at the back of the auditorium. It's early in the morning and, to my astonishment, vodka shots and chocolates are offered before the conference commences. Politely, I decline the offer, while the professors eagerly knock back one shot after another, all the while munching on chocolates. I catch a whiff of alcohol on the breath of the other speakers as I step onto the stage to deliver my speech. During my pauses for the translator to catch up, I can hear the professors engaged in their own conversations in the background. Clearly, they have little interest in what I have to say. Thirty minutes later, I step down from the podium, hoping that at least the students in the audience have gleaned something from my talk on government accounting.

As Vladimir and I hail a taxi to return to the hotel, he begins to share a bit about himself. He mentions that he was a student at St. Petersburg University, an institution notorious for producing FSB recruits. I can't help but wonder if his current position is a cover for clandestine activities, as he frequently disappears without his team being aware of his whereabouts. It will be a few years later that I learn from colleagues that he wasn't a secret agent after all, but rather had been running a separate business without permission from his employer.

Once Vladimir bids farewell, I make my way to the hotel bar to treat myself to some Russian champagne, locally known as shampanskoye. A couple of glasses later, I begin to unwind and realise that the day wasn't all that bad. And the adventure doesn't end there. I'm expecting Aliya, an old friend, to arrive any moment for an evening out in downtown Moscow.

Right on time, Aliya arrives and we hop into her Jaguar. She is so petite compared to the size of the car that one might mistake it for a driverless vehicle navigating the city streets. This is the second time I've been chauffeured by Aliya in Moscow, and I fondly remember her belief that parking rules do not apply to her. On one occasion, we boldly parked on Red Square without so much as a blink from the patrolling police officers. I'm delighted to see that she still possesses her rebellious spirit, along with her Jaguar.

As we gossip non-stop during our drive, we arrive at an exclusive jazz club nestled in an old warehouse. We enter through reinforced steel doors and descend a set of dimly lit steps into a smoky basement. Classic jazz fills the air as we indulge in a couple of glasses of burgundy. The clientele here are not your typical Russians, as I soon discover when Aliya introduces me to the Head of the Russian

Stock Exchange, who glides past our table, followed by several other figures from the Russian intelligentsia. I'm quite certain you won't find this club listed in any Moscow travel guide!

It's moments like these that remind me of the beauty of travel – the unexpected encounters, the hidden gems, and the tales we gather along the way. Moscow, with all its quirks and contrasts, never fails to surprise.

22 May 2010

As our car makes its way to the airport, I can't help but observe the pedestrians passing by. Their hunched shoulders and bowed heads seem to bear the weight of this harsh city environment. My mind begins to drift, drawing comparisons between Moscow and Kyiv, the capital of Ukraine, a city I have also had the opportunity to visit. The imposing Soviet architecture that dominates Moscow doesn't sit well with me, unlike Kyiv, which boasts a unique and diverse cityscape with fewer reminders of its Soviet past. And then there's the Dnipro River, a majestic waterway that outshines the poor, flat, and polluted Moskva River. But what truly sets Kyiv apart is its people – friendly, optimistic, and hopeful for the future. When it comes to the metro systems, Moscow's is undoubtedly impressive, with ornate designs and depictions of Soviet workers adorning its vast platforms. So, in my personal ranking, it's Kyiv: three, Moscow: one.

As we reach the departure terminal, a sense of relief washes over me. It's always nice to be heading home.

5: KENYA RHINO BABY

**London Heathrow Airport –
Jomo Kenyatta International Airport, Nairobi, Kenya**

Flight time: Eight hours and thirty minutes

22 February 2011

As I disembark from the plane, the warm embrace of the African sun greets me, instantly transporting me to daydreams of the vast Serengeti desert. In my mind's eye, I can already picture the majestic creatures of the savannah, the thrill of spotting the Big Five, and the vibrant local markets overflowing with exotic spices and musky scents. But alas, reality comes crashing back as I make my way through border control, reminding myself that this trip is for work.

My accommodation in Nairobi is nestled within the charming University Quarter, in a grand old colonial hotel that exudes a sense of timeless elegance. From its Art Nouveau veranda, I can leisurely observe the ebb and flow of life passing by. While I would love to venture out and explore the city, I've been cautioned against doing so alone. Nairobi's reputation for street crime is no laughing matter, with reports of violent incidents that can pose serious threats to personal safety. The ominous warnings from the UK Foreign Office about potential kidnapping and grenade attacks loom in the back of my mind. Despite the limitations, I can't deny that being confined to a luxurious five-star 'prison' for the next three days doesn't sound too shabby after all.

23 February 2011

Well, today's work assignment seems like a breeze compared to some of the challenges I've faced. All I have to do is present at a pan-African conference and mingle with fellow participants until my energy reserves are completely depleted. If everything goes according to plan, I'll have plenty of time to indulge in a sightseeing excursion tomorrow. And let me tell you, I'm more than willing to part ways with some cash for the opportunity of a lifetime – a safari experience that will knock an item off my 'bucket list'. After all, you only live once, right?

The conference venue is a sprawling circular hall adorned with vibrant flags representing the nations of southeast Africa. Balconies and multiple levels create a precarious scene as participants lean over to catch a glimpse of the action. The cacophony of noise is absolutely deafening. As I step onto the podium to deliver my presentation, I can't help but feel like an African general, rallying the troops in the face of adversity. It's no small feat to capture the attention of a crowd engrossed in technical discussions about accountancy. People are coming and going, cameras are flashing, mobile phones are ringing incessantly, and the constant hum of chatter fills the air. It's organised chaos, to say the least. But despite the hurdles, I manage to wrap up my presentation, eliciting a thunderous round of applause from the audience. My colleagues seem pleased and I'm relieved. With that task accomplished, I can now shift my focus to tomorrow's exciting extracurricular adventures. Let the fun begin!

24 February 2011

Before the break of dawn, my driver, Chane, arrives, and I discover that I'll be the sole passenger on this trip. Sensing my nervousness, Chane tries to put me at ease with a friendly introduction. "Hi, I'm Chane, your driver for the day. I'm married with two children." We embark on our journey, engaging in a conversation about his kids and the contrasting school systems in Kenya and the UK. It's a fascinating topic, especially considering that both systems have a mix of privately funded and public schools. However, in Kenya, only primary education is compulsory, while secondary education is optional, resulting in fewer children enrolling. This disparity poses challenges for social progress and economic growth, potentially hindering Kenya's transition from a developing to a developed country.

As we traverse the bustling streets, our conversation starts to wane, coinciding with our gradual departure from Nairobi's vast slums and informal settlements. In one particular area, a quarter of a million people live without access to running water, while proper sanitation remains a luxury. Chane skilfully makes a sharp left turn, leading us onto the old road to Lake Nakuru in the magnificent

Great Rift Valley. Stretching 6,000 miles from Jordan to Mozambique, the valley features a series of volcanic formations, many of which now lay dormant, and serves as a habitat for thousands of pink flamingos.

We wind our way through the valley on a narrow road, passing by a small chapel. Chane mentions that both the road and chapel were constructed by Italian prisoners of war during the Second World War. We make a brief stop, only to find the chapel closed. Continuing our journey, we encounter a congested road filled with gas-guzzling trucks racing past us from countries as far away as Uganda and Tanzania. The air grows heavy with the scent of diesel and road dust, but Chane adeptly manoeuvres our vehicle to avoid any potential collisions as these behemoths navigate treacherous bends. Just as we approach the Great Rift Valley, the dawn breaks, unveiling a majestic sight – a colossal golden sun rising over the ridge, revealing a inspiring view of an expansive landscape, where the ground seemingly disappears, extending for thousands of kilometres in every direction. We pass a sign on the right that reads 'This way to the Serengeti Desert'. A part of me yearns to turn right, but our journey continues straight ahead.

Arriving at Lake Nakuru, I'm overwhelmed by its beauty. Vast flocks of flamingos paint the lake pink as they bask in the sunlight. I spot rhinos, giraffes, and zebras adorning the rocky escarpments and the surrounding trees. We venture deeper into the national park in search of lions, but luck eludes us today.

Suddenly, the scent of smoke fills the air, and my eyes witness the devastating impact of fire on the landscape.

Gamekeepers armed with nothing more than branches struggle to contain the fires, looking defeated in their efforts. On the blackened horizon, a dazzle of zebras stand in fear, their once-vibrant surroundings now reduced to charred remnants. It's a heart-wrenching sight, one that we'll encounter all too frequently as the consequences of climate change escalate. Without decisive and unified action from governments, developing countries ill-equipped to confront the repercussions of rising global temperatures will face catastrophic consequences.

25 February 2011

With a heavy heart, I board the plane, holding onto a treasured possession – an oversized wooden giraffe – as a reminder of my connection to wildlife. Thoughts of my adopted rhino, Kyjanu, fill my mind, knowing that he has been relocated to Nairobi's national park. Regret seeps in as I realise that I've missed the chance to meet my beloved baby boy. A tear wells up in my eye as the plane takes off from the runway, carrying me away from the opportunity to see Kyjanu in person.

6. WHO WOULD SAY NO TO THE CARIBBEAN?

**London Gatwick Airport –
Norman Manley International Airport, Kingston, Jamaica**

Flight time: Nine hours and forty minutes

6 March 2011

I arrive at Gatwick airport, ready for my evening flight and go through the familiar routine of removing my shoes and belt, placing my laptop and mobile phones in a tray, and emptying my pockets of money and keys. Just as I think I'm done with security, the highly sensitive alarm is triggered by the metal underwiring in my bra. Oh, the joys of modern airport security!

After finally making it through security control, I discover that my flight is delayed by several hours, which is extremely frustrating. While I'm thrilled to be heading to the Caribbean for the second time in my life (the first being a trip to Cuba), the excitement is slightly dampened by the prospect of sitting in soulless airport lounges for hours on end. There's only so much coffee one can drink to pass the time. But hey, at least I'll have plenty of time to catch up on some reading or people-watching.

7 March 2011

It's late afternoon when our plane touches down at Norman Manley Airport, and my tired eyes scan the crowd for George, the designated driver assigned by the local office. Spotting a young man approaching and calling my name, I assume it must be him. Fatigue weighs me down, and without a second thought, I hop into the car.

The road from the airport to Kingston's city centre is infamous for its carjackings and robberies, but I've been assured that George will ensure my safe arrival at the hotel. As we're en route his phone rings and he casually tosses it back to me. To my surprise, the caller on the line is George himself!

My brain takes a moment to process the situation. Oh no, who is driving this car? Is this a kidnapping? Panic sets in as I consider my options, contemplating jumping out or calling for help. Sensing my distress, George quickly intervenes, his voice oozing with reassurance. "Don't worry, you haven't been kidnapped," he says. "Bernard, your driver, is a family friend. I apologise for not being able to meet you at the airport." With a sigh of relief, we continue on our way, and I finally arrive safely at the hotel, grateful that it was all just a misunderstanding.

8 March 2011

As I make my way to breakfast, I notice that the dining room is sparsely populated, with only a few individuals present. Amongst them is a middle-aged businessman, his greying hair and paunch indicating a life of privilege, and a young Jamaican woman dressed in revealing attire, exchanging money discreetly under the table. They sit in silence, their transaction concealed from prying eyes.

While most visits to Jamaica are trouble-free for tourists and business travellers, occasional reports of crime and violence, especially in Kingston, linger in the back of my mind. Seeking reassurance, I approach the receptionist and enquire about safety outside the hotel. To my surprise, they assure me that it's safe to walk around, contradicting the warnings I received from my employer. However, I recall a recent incident where a colleague was mugged at knifepoint just outside this very hotel. With this knowledge, I decide it's best to remain within the secure confines of the hotel for the time being.

As I sit by the pool, I delve into a travel guide about Kingston, discovering that it boasts one of the highest murder rates amongst urban areas in Jamaica. The drug economy looms large, constituting a significant portion of the country's GDP and posing a grave challenge to governance. The book further informs me about the prevalence of gang violence, shootings, and robberies in the city and surrounding neighbourhoods. Adding to the disconcerting information, I learn that the hotel I'm staying in is reputedly owned by an infamous drug lord. Given these circumstances, I find solace in the secure compound of the hotel, sipping on a non-alcoholic piña colada while attending to work emails.

Despite the alarming crime statistics and hardships faced by the Jamaican people, I hold onto the belief that I am about to experience the warmth and hospitality of individuals with an exceptional sense of humour.

9 March 2011

This morning, Monica, my colleague from Trinidad, arrives in Jamaica. It's a joyous reunion as we haven't worked together since our time in Belize. Together, we head off to conduct a seminar with government officials on the pressing topic of anti-fraud and corruption. As we step into the foyer, I finally meet George in person, and he offers another apology for not being able to pick me up at the airport yesterday.

To our surprise, the number of participants for the seminar has skyrocketed from the expected forty to a staggering 250. Our meticulously planned agenda for a small group setting goes out the window in an instant. We have no choice but to think on our feet and adapt. Thankfully, the participants are enthusiastic and engaged, bringing forth a wealth of ideas and experiences on combating fraud. The room buzzes with excitement as individuals form a winding queue to share their insights at the microphone. Amidst the serious business of addressing anti-fraud and corruption in Jamaica, I can't help but appreciate the dark sense of humour exhibited by the participants, who jokingly mention the risks involved, including the possibility of anti-corruption enforcers being shot dead.

Undeterred by a tropical storm, we swiftly move on to our next seminar, which focuses on ethics with a group of finance students. The heavy rainfall causes delays as they trickle into the classroom, hindered by the city's traffic coming to a standstill. Once we finally commence, it becomes evident that the students are struggling to grasp even the most fundamental concepts of ethics.

It's a challenge that, if left unaddressed, could impede their progress in their future careers in accountancy. As the tropical storm gradually subsides, we retreat to the comfort of our accommodation, affectionately dubbed the 'drug baron's' quarters, where we reward ourselves with well-deserved alcoholic piña coladas.

10 March 2011

Yesterday evening, we embarked on a flight to Port of Spain, Trinidad, to continue our schedule of seminars and meetings. As we made our way to the hotel, our driver made an unexpected stop at a popular viewpoint, where he regaled us with chilling stories of shootings and muggings that had occurred at that very spot. His anecdote included a gruesome tale of a couple being shot at point-blank range while sharing a romantic moment. It was certainly an unsettling introduction to Trinidad.

In contrast to Jamaica, Trinidad boasts significant wealth and development, largely attributed to its petroleum-based economy. However, I soon learn from Monica that the country faces similar crime levels and serves as a transit route for illegal drugs. Once again, I'm advised to stay within the confines of the hotel grounds, but if I go out not to venture left, as it leads to areas rife with drug-related activities.

Today, the atmosphere at the seminar is markedly different from our experience in Jamaica. The participants seem more reserved and less inclined to ask questions or actively engage in the discussion. Monica and I find ourselves working twice as hard to maintain the momentum. People appear fatigued, and the room lacks the same sense of humour that we encountered in Jamaica. As the hours drag on and weariness sets in, Monica delivers a closing speech that rivals that of a seasoned political leader. It comes as no surprise to me, knowing that she had once contemplated a career in politics. Her impassioned address leaves the participants with a final ethical question to ponder: Is it ethical to drive in the bus lane when you know you're not supposed to do it? Finally, the room erupts in uproarious laughter, a clear indication that many have likely succumbed to that temptation at some point.

7. SOUTH KOREA – KIMCHI AND THE KITCHEN BATTLE

**London Heathrow Airport –
Incheon International Airport, Seoul, South Korea**

Flight time: Ten hours and fifty-five minutes

19 May 2011

It's 9pm and I find myself navigating the lively streets of downtown Seoul with my colleague, Anna. The bustling atmosphere resembles a busy Saturday afternoon on London's Oxford Street. It is teeming with vendors selling cheap clothes, counterfeit watches, and fake handbags.

Anna is craving a beef dish, whilst I hope to find a vegetarian option. However, our search for a restaurant catering to my dietary preference proves challenging as we peruse menu after menu. It seems that the choices are limited to either beef or pork, or pork or beef. Anna translates the menus for me, pointing out dishes like *gopchang* (beef or pork intestines), *sundae* (blood pudding), and *gamjatang* (pork bone soup). I'm not having much luck finding a vegetarian-friendly establishment, and Anna's hunger is turning into irritability. Finally, we stumble upon a restaurant that offers omelettes.

As we sit down and examine the menu, Anna insists that I try the national dish, kimchi. This traditional dish, dating back 2,000 years, involves salting and fermenting cabbage with a mixture of pepper, garlic, ginger, and spring onions. Nowadays, kimchi can be found in restaurants and food markets worldwide, including the UK.

However, when our order arrives, there is no beef in sight, only a plate piled high with vegetables. This is an absolute catastrophe for Anna, as she is adamant about having her beef and refuses to settle for anything less. She transforms into a woman possessed and calls the waiter over, engaging in a heated argument with him in rapid-fire Korean. The other diners find the spectacle amusing, as the commotion escalates with both Anna and the waiter shouting at the top of their lungs. It's my first time meeting Anna, and her assertiveness is both surprising and slightly intimidating.

Determined not to back down, Anna brushes the waiter aside and storms into the kitchen to confront the cook, who happens to be Chinese. She effortlessly switches from Korean to Mandarin, further escalating the confrontation as the chef brandishes a knife. The entire restaurant falls into complete silence, with all eyes fixed on Anna and the cook. In Mandarin, the chef informs Anna that there is no beef left and she will have to make do with the vegetables. Anna retorts, "Go and find some beef then, this is appalling service." She has the final word, but her energy dwindles, and she reluctantly returns to the table to begrudgingly eat the vegetarian meal. Sulking, she begins to tuck into the vegetables, and the rest of the restaurant resumes their conversations.

To our surprise and my relief, ten minutes later, a waiter scurries over with a plate of beef. Anna's demeanour swiftly changes, and normalcy is restored. She happily indulges in her long-awaited beef, and I, on the other hand, pass on the kimchi after giving it a try and deciding it's not to my taste.

20 May 2011

As I wander through the streets of Seoul, deep in thought about the previous day's finance conference, I reflect on its lack of diversity. Being the only female speaker and one of the few women in attendance highlighted the ongoing struggle for gender equality in the field of public financial management. While progress has been made over the years, senior positions remain predominantly occupied by men, and the lack of data on women in leadership positions further obscures the issue. It was this experience that would eventually inspire me to establish Public Finance by Women, an organisation advocating for gender equality and the collection of relevant data.

During the final group photograph, I overheard a comment referencing the phrase 'you are the only woman in the village', a line borrowed from the UK comedy sketch show, *Little Britain*. Although intended as a light-hearted remark, it underscored the underrepresentation of women not only at the conference but also in society at large. Gender equality still has a long way to go.

Continuing my walk, I am struck by the overwhelming presence of concrete high-rise buildings. Seoul's skyline is dominated by towering structures, showcasing a level of brutalist architecture unlike anything I have seen before. The streets teem with people, and the population appears homogenous compared to other cities I have visited. Seeking respite, I enter the gates of Gyeongbokgung Palace, a historic site built in 1395 and now a UNESCO World Heritage site. I observe the mesmerising changing of the guard ceremony and find solace in the enchanting palace gardens, taking a moment to rest my tired feet.

Venturing further, I meander through boulevards flanked by skyscrapers until I reach narrow lanes filled with small businesses and kiosks selling various goods. Suddenly, I am captivated by a vibrant display of brightly coloured lanterns adorning the exterior of a Buddhist temple. Shades of red, pink, orange, blue, and green create a stunning spectacle. The sweet scent of incense fills the air, accompanied by the gentle sound of chanting emanating from within the temple. I find myself irresistibly drawn to this place of tranquillity and spirituality. As I sit sipping my coffee, I observe worshippers coming and going, witnessing their interactions, and feeling a profound sense of peace. The temple's allure is so intoxicating that I even contemplate the idea of converting to Buddhism, unable to tear myself away from its enchanting ambiance.

21 May 2011

As I explore the city square, I stumble upon an open-air rock concert, featuring songs from rock groups that I don't recognise. A university professor approaches me and enquires about my visit to Seoul, also asking if I am aware of the purpose of the concert. While I had a hunch that it was connected to a politician or political party, it was merely a guess. The professor proceeds to inform me that the concert is a tribute to Roh Moo-Hyun, a former president of South Korea. He explains that Roh Moo-Hyun allegedly died by leaping from a hill in 2009 following a bribery scandal, and the people attending the concert are his supporters who refuse to believe that he committed a crime. Instead, they are celebrating his life and embracing a conspiracy theory that he was set up by another political party.

Feeling a sense of intrigue, I depart from the vigil, leaving behind thousands of burning candles swaying in the darkness. Upon returning to the hotel, I decide to conduct further research on Roh Moo-Hyun. I delve into the details of his life and discover that South Korea has the second highest suicide rate in the world amongst developed nations. This sobering information sheds light on the profound social and mental health issues that the country faces.

Reflecting on this unexpected encounter and the information I have acquired, I contemplate the complexities of political scandals, the power of conspiracy theories, and the impact they can have on society.

22 May 2011

As I celebrate my birthday heading back home, the eleven-hour flight is made more enjoyable with unlimited champagne. Time seems to pass swiftly as I reflect on my experiences and the journey I have undertaken. Cheers to more delightful and interesting memories!

8. A WORLD CUP QUALIFIER IN BUENOS AIRES!

**London Heathrow Airport –
Ezeiza International Airport, Buenos Aires, Argentina**

Flight Time: Fourteen hours and eighteen minutes

8 November 2011

I arrive in Buenos Aires after a long and tiring flight, and I'm greeted by conference officials. However, the story shared by a fellow traveller about a disturbing incident where a businessman woke up in a hotel room to find he had his kidney extracted leaves me with a sense of unease. It serves as a reminder to be cautious and vigilant, especially when travelling to unfamiliar places. While such occurrences are rare, it's important to prioritise personal safety and take necessary precautions. Nonetheless, I can still look forward to exploring the vibrant city of Buenos Aires and making the most of my time here.

9 November 2011

It can be quite challenging to navigate through a conference while dealing with jet lag, but I manage to deliver my presentation on sustainability reporting to the audience of environmental auditors. It's ironic that I receive a heavy paperweight as a gift, considering the environmental focus of the event and the carbon footprint associated with transporting it back to the UK. Sometimes these small ironies remind us of the complexities and contradictions we encounter in our daily lives, even when striving to promote sustainability.

As I attend the conference and listen to various presentations about auditing environmental issues, I find the initiatives and topics discussed to be intriguing and important. However, one presentation stands out as peculiar when they mention including a polar bear in their scope of review. Given that polar bears are not native to Africa, it seems like an odd example to cite in the context of African flora and fauna protection. It may have been a simple mistake or a miscommunication, but it's caught my attention and seemed out of place. Nonetheless, the overall focus on protecting biodiversity and natural resources in Africa remains crucial for sustainable development.

10 November 2011

You won't believe the thrilling news I discovered at the conference today! Tomorrow evening, there's a FIFA World Cup qualifier match between Argentina and Bolivia, and it's happening at none other than the iconic River Plate Stadium, also known as El Monumental. This stadium, which was built back in 1938, can hold over 70,000 passionate fans and serves as the home ground for Club Atlético River Plate, one of Argentina's most beloved football clubs. Of course, Boca Juniors is the other legendary club that captures the hearts of the Argentine people. It's no exaggeration to say that more than half of all Argentinians are die-hard supporters of one of these two teams.

But let's get back to the match at hand. The River Plate Stadium is famous for hosting the 1978 FIFA World Cup, a momentous occasion in football history where the home team triumphed. And guess what? I've managed to get tickets for the game! I quickly sent a WhatsApp message to Aidan, my partner who will be joining me on this trip, and exclaimed, "I've got tickets!" This is going to be an unforgettable experience, and Aidan will be over the moon when he hears the news. The electric atmosphere, the roaring crowd, and the pulsating energy of Argentine football await us. I can hardly contain my excitement!

11 November 2011

As we made our way to the northeast section of the River Plate stadium, a mix of excitement and unease filled the air. This stadium had gained a notorious reputation for fan riots in the past, leaving players needing police protection and the stadium in disarray. The memories of smashed shop windows and injured individuals lingered, and the heavy presence of riot police armed with automatic weapons reminded us of the potential dangers. The atmosphere is simultaneously thrilling and terrifying.

Entering the stadium, we were greeted by the thunderous beats of drums and the sight of brightly coloured flares illuminating the pitch. The bowl-shaped arena was closely monitored by heavily armed police patrolling the aisles. Initially, I found myself seated next to a concrete column, feeling trapped and overwhelmed by claustrophobia. I quickly moved to an empty seat, seeking relief from my fear.

The stadium was adorned with blue and white flags, a sea of support for the home team. And then, in a moment of disbelief, Lionel Messi, the iconic footballer and one of the greatest players in the world, stepped onto the pitch. It was a dream come true, ticking off boxes for River Plate, Messi, and a FIFA World Cup qualifier. We were incredibly fortunate to witness such a momentous event.

The *Buenos Aires Herald* had reported that Alejandro Sabella, the new coach, would deploy three attackers in this game.[1] The pressure on the Argentine team was immense. Losing would be seen as a national tragedy. However, the game itself proved to be underwhelming. The Argentine team appeared lacklustre, with Messi carrying much of the burden. The match ended with a 1-0 victory for Argentina, but it was far from a convincing performance. The following day's press would surely have a field day with their analysis.

Thankfully, we didn't witness angry fans tearing up seats or violence spilling out onto the streets, nor the use of tear gas or water cannons. Although the atmosphere in the stadium was at times edgy, I was relieved to reunite with our tour guide who swiftly led us back to the safety of the bus.

---

1   *"Sabella searches for balance", Buenos Aires Herald, 10 November 2011*

12 November 2011

We wake up to the harsh critiques of the Argentine team in the early morning papers, particularly *La Nacion Deportivo*. The headline, '*Lamento Argentino*' reflects the disappointment, and individual players are singled out as major let-downs.[2] It's clear that the performance has not been well received.

Buenos Aires truly is a captivating city, boasting a population of over 3 million people. It stands out as the most racially diverse city in South America. With my work commitments behind me, Aidan and I embark on exploring the city before venturing to the enchanting region of Patagonia. As we wander through Buenos Aires, the streets are alive with activity, and the grand, faded colonial architecture captivates us at every corner. Even in the more traditional neighbourhoods like Palermo, the largest barrio, we're treated to colourful murals, created by local artists, showcasing the city's vibrant street art scene. We pause for a coffee break, savouring the ambiance before continuing our journey.

Our path leads us through public spaces adorned with magnificent jacaranda trees in full purple bloom. We stumble upon expansive areas filled with every imaginable breed of dog. It's no secret that Buenos Aires has a profound love for canines, with over 500,000 pet dogs residing in the city. Dog-watching becomes just as popular as people-watching, and it provides us with endless entertainment. We marvel at the '*Paseadores de Perros*', the dog walkers who skilfully handle groups of

---

2   "*Lamento Argentino*", *La Nacion Deportivo*, 12 November 2011

five to twenty dogs at a time. The pet shops, extravagantly adorned, offer a vast array of pet products, rivalling even the most expensive children's clothing stores. Our leisurely stroll is interrupted by thoughts of the national strike, wondering if our flights will operate as scheduled. To our dismay, we are flying with an airline known for its dismal reputation, living up to its title as the worst airline in the world.

13 November 2011

Although today marks the beginning of our holiday, it would be remiss not to mention the enchanting region of Patagonia. However, our journey to this picturesque destination was marred by unexpected chaos at the airport. As we eagerly awaited our flight, we were met with a disheartening sight; no flight details were displayed on the electronic boards, and there were no airport staff available to provide any information. Despite having completed the check-in process, we found ourselves in a state of anxious anticipation, desperately waiting for some news.

Five long hours elapsed before a glimmer of hope appeared: the flight number miraculously materialised on the electronic display. We hastily made our way to the gate, racing against time to board the final and only flight bound for El Calafate in South Patagonia.

14 November 2011

El Calafate, a town nestled on the edge of the Southern Patagonian Ice Field, serves as our gateway to the majestic

Perito Moreno Glacier. Today, we embark on a journey to witness this natural wonder first hand. Against a backdrop of rugged, towering mountains, a vast expanse of glistening white and blue ice captivates our senses with its beauty.

As we draw closer to the glacier, we are greeted by the awe-inspiring sight of massive blocks of ice calving and plunging into the nearby lake. Boarding a small boat, we brave the fierce gusts of wind, battling against the elements as we venture closer to the face of the glacier. On deck, the gusts make it difficult to maintain our balance, but the sheer spectacle before us keeps us grounded. In the midst of the howling wind, we are serenaded by the haunting sounds of cracking ice, reminiscent of distant gunshots, as mighty chunks of ice break free and crash into the lake just metres away from our boat.

This is just one of the many glaciers we will explore within the national park. It is disconcerting to realise that these glaciers, which hold such natural magnificence, are rapidly disappearing due to the alarming pace of melting. The urgency to address climate change and protect these delicate ecosystems becomes all the more palpable, as we witness first-hand the inevitable loss that awaits us in our lifetime.

15 November 2011

This morning we embark on a leisurely stroll through the Nimez Lagoon Nature Reserve, located just outside of El Calafate. This sanctuary is renowned for its rich biodiversity, particularly its avian inhabitants, with a reported eighty species of birds finding solace in its wetlands, scrubs, and lake shore.

Little did I know that I would soon have a thrilling encounter with a peregrine falcon. Coming face to face with this majestic bird, I am momentarily frozen as it clutches its fresh prey. Sensing the intensity of the moment, I cautiously move away, but my respite is short-lived. Within seconds, I feel a presence circling above me, and as I raise my eyes, the falcon hurtles down towards me with lightning speed. Fear engulfs me, and I instinctively scream, raising my backpack as a feeble defence. It hesitates, retracting its menacing talons, only to resume its circling motion. Panic takes over as it makes a second dive towards me. I stumble backward, brandishing my backpack in the air, desperately warding off the impending threat. "Aidan, help!" I cry out, and he rushes to my aid, urging me to seek shelter.

With adrenaline coursing through our veins, we retreat to safety, taking refuge in a nearby bird hide. Trembling and shaken, we are hesitant to venture outside for what feels like an eternity. Eventually, cautiously following another visitor's lead, we emerge from our sanctuary, still on edge. Later, we learn that this bird has a reputation for its aggressive behaviour, having targeted unsuspecting visitors on previous occasions. It seems my inadvertent proximity to its nest triggered its protective instincts.

As we bid farewell to the reserve, we witness a spectacle that further reinforces nature's dominion. Two screeching birds chase a hapless dog, swooping low and reclaiming their territory. In this realm, nature reigns supreme, reminding us of its untamed power and our humble place within it.

16 November 2011

After the harrowing bird encounter, we embark on a five-hour journey to El Chaltén, a village nestled within the Los Glaciares National Park. Our goal is to trek the trails that wind through the Andes, offering spectacular views of Cerro Torre and Mount Fitz Roy. The Andes, spanning across seven countries, stands as the longest mountain range in the world, and here in El Chaltén we are surrounded by its majestic peaks and vastness. The village itself exudes a rustic charm, reminiscent of a Wild West border town, with its windswept streets and free-roaming stray dogs.

Setting off on a gloriously sunny day, we cross a bridge over the Blanco River and begin our trek towards a verdant hillside forest. As we follow the meandering river, gradually ascending, we enter the gateway to the National Park. Despite my lingering nervousness triggered by the peregrine falcon attack, I find solace in the presence of Tom, our experienced guide and protector. He assures me that such encounters are not uncommon and shares stories of his own close confrontations, including one with a puma along this very trail. My senses now alert, I scan the surroundings for any signs of the elusive big cats. Amidst our conversations, the western peaks of the mountains start to peek through the dense foliage, with Mount Fitz Roy commanding attention with its imposing north face.

After a couple of hours of trekking, we pause at a viewpoint that offers a glimpse of the Piedras Blancas Glacier, nestled within a steep-walled valley. While it may not be the most spectacular glacier we have come across during our journey, it serves as a testament to the natural wonders that surround us. We have been walking for quite some time, and now faced with a decision, we contemplate whether to retrace our steps or continue on the longer route towards Camp Poincenot, the base camp for climbing the Fitz Roy mountain range. Opting for the latter, we forge ahead, traversing through scrubby beech forests, marshlands, babbling brooks, and rocky terrain, all the while being embraced by the jagged silhouette of Fitz Roy. The path leads us deeper into the valley, eventually guiding us back towards El Chaltén.

Descending into the village, we believe our Patagonian adventure is coming to a close. However, fate has a different plan as we discover that the airline strike has left us stranded, unable to depart this enchanting land. It seems that Patagonia has chosen to hold us captive a little longer, extending our journey and beckoning us to explore further.

9 BANGLADESHI FEVER

London Heathrow Airport –
Dubai International Airport, UAE –
Hazrat Shahjalal International Airport, Dhaka, Bangladesh

Flight Time: Thirteen hours plus/minus an hour for changeover

25 February 2012

As I embark on another journey with Emirates, I can't help but indulge in the small pleasure of witnessing my neighbour's curtains twitch as a sleek, black Mercedes Benz glides onto the road. The presence of a chauffeur, gracefully assisting with the loading of luggage, adds an air of sophistication to the surroundings. Amongst the assortment of vehicles, it undoubtedly stands out as the poshest car in sight.

Our familiar destination is Heathrow Airport, a place I find myself returning to once again. It's a hub of bustling activity, a melting pot of travellers from all walks of life. As I make my way through the airport, I'm already plagued by sniffles and a sore throat, an unfortunate omen before a long-haul flight to a developing country. The very thought of its noted high pollution levels sends a shiver down my spine, knowing that it will likely exacerbate my throat discomfort. And to add to the worry, the availability of medicines in this destination might prove to be a challenge.

Despite these concerns weighing on my mind, I'm momentarily distracted during the first leg of the journey to Dubai. An air steward mistakes me for a TV news presenter, and while I'm flattered, he keeps the mystery alive by not revealing who he has confused me with. With a mischievous smile, he quips, "You must get asked this question a lot." His unexpected remark breaks the tension, eliciting a genuine burst of laughter from me, momentarily overshadowing the worries that loom ahead.

26 February 2012

After enduring a thirteen-hour flight, I finally touch down at Hazrat Shahjalal International Airport, only to be greeted by scenes of absolute chaos. The queue to pay for a visa winds its way through the arrivals room, out the door, and onto the runway, resembling a scene from *The Great Escape*. Jumping the queue by offering a small bribe is an option for some, but my principles prevent me from succumbing to such practices. It takes an eternity to navigate through the congested traffic and leave the airport behind.

As we make our way through the bustling streets of Dhaka, I'm immediately struck by the alarming state of the environment. Thick black smoke billows from the exhaust pipes of lorries, mingling with the gritty air that is filled with dust from the uneven roads. Coughing fits ensue as the combination of dirty diesel and swirling dust permeates the car's interior. The roads themselves are riddled with bumps and remain partially unfinished, with half-built highways towering above buildings on concrete stilts. Metal rods jut out from sections of cement, showcasing the premature deterioration of the infrastructure. During my flight, I had read about Dhaka's poor urban air quality, and now I can feel the effects first hand as my eyes sting and my sore throat worsens.

Dhaka, with its staggering population of 10 million, epitomises overcrowding. Bangladesh as a whole, with its total population of 171 million, ranks amongst the most populous countries in the world. It's ironic that while real estate prices soar, the majority of people live below the poverty line. Observing the people on the streets, it's evident that they are struggling to navigate their daily lives. A dilapidated bus, resembling a bashed-up tin can on tractor wheels, passes by, belching out noxious fumes. Packed with passengers, the bus is a grim reminder of the challenges faced by the city's inhabitants. I witness a young girl leaning out of the window, her head hanging over the glassless frame as she vomits onto the rubbish-strewn roadside. My initial impression of Dhaka is far from positive, but I remain hopeful that my perspective will change over time.

I arrive at one of the few five-star hotels in Dhaka, where stringent security measures are in place. Armed guards patrol every entrance, even the rooftops, underscoring the need for heightened safety precautions. However, my enthusiasm for a comfortable stay wanes as I'm assigned a smoker's room. The air is heavy with the stale stench of nicotine, leaving me no choice but to request a room change. Despite the switch to a non-smoker's room, the lingering smell of cigarettes permeates the walls, seemingly ingrained into the very plaster. To make matters worse, the antiquated air conditioning system leaves me wondering what airborne ailments might be circulating within the confines of my room. The temperature is frigid, exacerbating my hoarseness and discomfort.

Sitting by the window, gazing out at the high perimeter walls guarded by armed personnel, my exhaustion from jet lag weighs heavily upon me. Beyond those walls, sprawling wastelands and slums stretch into the distance, a stark reminder of the vast disparities that exist in this city. Overwhelmed by a sense of guilt, I grapple with the knowledge that I have the luxury of staying in a comfortable hotel, while countless individuals struggle to survive just beyond its confines.

27 February 2012

OMG, I wake up with a throat on fire and an empty fridge that once held bottles of water. My energy is already depleted before the day even begins.

Parveen waits for me in the lobby, and together we head to Dhaka University to address a diverse audience of students, professors, and politicians on the crucial topic of improving financial management in low-income countries. I feel a deep sense of humility speaking at this prestigious institution known for its strong political science program and the illustrious politicians it has produced. We arrive early and sit in the car, but the driver leaves the engine running, filling the interior with petrol fumes as he lights up a cigarette nearby. Gasping for fresh air, I quickly move away until we are invited inside the university.

My presentation was initially scheduled for just half an hour, but to my surprise it stretches on for two hours as more speakers seemingly emerge out of thin air. We endure two electric power cuts that plunge us into darkness while enduring a steady stream of political speeches focused on the Bangladeshi economy. Unbelievably, the prime minister's brother-in-law decides to make an appearance and contribute to the discussions. Although most of his speech is delivered in English, he occasionally switches to Bengali, possibly to make fewer flattering remarks about my talk or the UK's financial management track record. Knowing the sensitivity of government officials to any form of criticism, I am careful with my words and actions. Unfortunately, the country's record on political freedoms and human rights has been deteriorating, as reported by campaigners.

Despite resting my voice and hydrating the previous evening, I find my voice increasingly husky. I struggle through the final meeting with government officials and business leaders, including the UK ambassador. Suddenly, in the midst of the discussion, my voice gives out, and I croak. Thankfully, Parveen steps in, attempting to cover what she believes I was going to say. Eventually, we make our way back to the hotel, manoeuvring through the streets filled with elaborately designed rickshaws before getting stuck in seemingly endless gridlocked traffic.

My misery deepens as I've lost my voice, feel worn down by a cold, and suffer from lingering jet lag. Unable to sleep, I resort to consuming chocolate, sucking on lozenges, and once again emptying the fridge of water. I grab a local newspaper, the *PABNA*, and stumble upon an article about a tragic incident where a schoolboy dies after being bitten by a dog.[3]

The article recounts the story of a schoolboy who was bitten by a dog on Monday in Baghail Purbopara, Iswardi Upazila. Tragically, he passed away shortly afterwards having dismissed the small wound and not informed his family about the incident. He had experienced chills and sickness and was rushed to Rajshahi Medical College Hospital, where he was found to be severely infected. Sadly, they sent him back home, stating that it was too late to save him.

Another heart-wrenching story featured in *The Independent* highlights the eviction of slum dwellers, often referred to as 'the have-nots', from railway land they had occupied for over three decades.[4] Despite their efforts to form a human chain and submit a memorandum to the prime minister, pleading for their case to be heard, the railway company continued to press ahead with eviction.

Having read these distressing stories, my mood sinks even lower, prompting me to reach out to Aidan back home to share my troubles. He's at a loss for words, unable to provide solace except for mentioning that Officer Dibble, our cat, is missing me.

---

3   *"Schoolboy dies after dog bites"*, PABNA, 24 February 2012
4   *"Slum dwellers demand rehabilitation first"*, The Independent, 25 February 2012

28 February 2012

With the throat lozenges finally providing some relief, I await my turn to present at the conference. However, things take a turn for the worse when the panel chair, who happens to be Bangladeshi, begins by rudely addressing a speaker from Pakistan, accusing him of being an imposter. It turns out the Pakistani speaker is a last-minute replacement for someone who fell ill. To add insult to injury, the chair scolds his own adviser for not informing him of the speaker change, causing the adviser to quickly make an exit. As if that weren't enough, he abruptly cuts short the Pakistani speaker's presentation, declaring, "Time is up," before turning to me and asking if the throat lozenges I'm using are recreational drugs. The uncomfortable atmosphere weighs heavily on the room, with colleagues shifting uncomfortably in their seats.

During the break, I find myself seeking refuge in the ladies' restroom to avoid an elderly freedom fighter who has been following me for the past two days. I've grown tired of his constant presence. While I understand the significance of the freedom fighters who fought for Bangladesh's liberation in 1971 during the brief conflict between India and Pakistan, establishing Bangladesh from the former East Pakistan territory, I'll never know if this individual genuinely participated in the war.

Fortunately, there is a silver lining to the conference as we are featured in the *Daily Star* newspaper, advocating for government bodies to improve their financial transparency.

Finally, the day comes to an end and we are herded together for an evening sightseeing tour of Dhaka. However, it turns out to be a disappointment as the car we are in has blacked-out windows, and it's night-time, making it impossible to see anything. Perhaps this was all part of the plan, considering there isn't much of interest to see besides the stagnant traffic. On a positive note, during our return to the hotel, I manage to come across a watercolour painting depicting the chaotic streets of old Dhaka. It still hangs in my home today and remains one of my favourite pieces.

10. SRI LANKA SEA BREEZE

Hazrat Shahjalal International Airport, Dhaka, Bangladesh –
New Delhi Airport, India –
Bandaranaike International Airport, Colombo, Sri Lanka

Flight time: Eight hours – it would have been three hours with a direct flight!

29 February 2012

I find myself pondering the decision to book a flight with a layover in New Delhi instead of opting for a direct flight to Colombo, which would have significantly reduced my travel time. It's currently 1am, and I'm seated in Hazrat Shahjalal International Airport engaging in a conversation with a buyer from a prominent high-street store about the garments industry and the quality of women's underwear. She explains how the garment industry plays a vital role in Bangladesh's journey out of poverty, transforming the country into one of the world's leading clothing exporters. Through manufacturing clothes for renowned brands, the industry has brought about economic and social empowerment.

The buyer has just returned from a factory in the Chittagong region of Bangladesh, where she conducted a health and safety audit. Such audits have become necessary following numerous scandals involving sweatshops and child labour. In this mediocre airport lounge masquerading as a business-class area, we find ourselves tired and patiently waiting for our flights to be called. I can only imagine how exhausting it must be for her, making this journey to Bangladesh twice a month.

And here I am once again, rushing at full speed across the concourse of Indira Gandhi International Airport in New Delhi to catch my connecting flight to Colombo. The security check-in line is long, and in my desperation to make the flight I find myself pushing my way through, much to the annoyance of fellow passengers. My heart pounds with anxiety, fearing that I might miss the flight and be stranded at the airport for another twenty-four hours without an Indian visa, unable to leave the confines of the airport.

As the flight touches down in Sri Lanka, I step out of the airport and am greeted by a warm sea breeze that gently caresses my face. Almost magically, my mood begins to shift. To my delight, I receive a hotel upgrade to a luxurious business suite with a stunning view of the sea. Even my sore throat shows signs of improvement. It's time to call Aidan and let him know that I'm feeling much better now. Unsurprisingly, he responds with little surprise when I describe my current surroundings.

Tomorrow, I'm looking forward to a well-deserved break.

1 March 2012

Abida my colleague and I meet for breakfast, excited to spend some time together and get to know each other.

It's my first encounter with her, and I'm intrigued by her chic and vibrant *salwar* outfit. These *salwars* can be quite expensive, with some costing thousands of pounds. While it looks fantastic on her, I know that pulling off the Bollywood look is a challenge for most Western women, myself included.

Soon, we hop into a chauffeur-driven car and embark on a journey to Galle in the southwest of Sri Lanka. The road is bustling with activity, as we navigate our way through cows and motorbikes. Cricket enthusiasts would recognise Galle as the venue for test matches between Sri Lanka and England. As I gaze out of the car window, I can't help but notice the abandoned buildings, mere shells of their former selves. The remnants are a haunting reminder of the devastating tsunami that struck in December 2004, claiming the lives of over 30,000 people. Along the roadside, I spot scattered ruins of houses, left behind by families who either moved to higher ground or tragically never returned. It's a beautiful drive, yet emotionally poignant.

Throughout the journey, Abida and I engage in lively gossip about office politics and life in general, with occasional pauses for Abida to pray towards Mecca. As girls often do, we make a stop for lunch and indulge in some shopping until we can't shop anymore. The old Portuguese town of Galle enchants us with its historical fort ramparts dating back to 1588 and its winding narrow streets that we navigate with delight. It's a fabulous day, and with our precious collection of gems and intricately embroidered throws, we make our way back to Colombo, filled with memories of our adventure.

2 March 2012

The first of our two working days in Colombo rushes by in a whirlwind of meetings with the university and the Ministry of Finance. Everything is going according to plan, except for one minor hiccup: the conference venue. It's situated right in the middle of a hotel atrium, and the constant stream of noisy tourists passing by becomes a major distraction for both participants and speakers. We struggle to maintain focus as heads turn to catch glimpses of the commotion. To make matters worse, the incessant sound of running water from an artificial waterfall creates a constant background noise that makes it difficult for me to be heard during my speech.

Aside from these challenges, there was one particularly memorable story shared by a guest today. He recounted an incident when he returned to his hotel room to find the cleaner brushing her teeth with his toothbrush. The sheer horror of the situation prompts me to retreat to my own room immediately, where I secure my electric toothbrush in the safe. From this day forward, it becomes an eccentric habit of mine to safeguard my toothbrush no matter where I travel in the world. One can never be too cautious!

As the day draws to a close, I reflect on the events of today and realise that despite the usual business activities, there were no particularly unusual occurrences. However, the toothbrush incident serves as a reminder to expect the unexpected in every situation. With that thought in mind, I retire to my room, ready to face the next day's adventures with a touch of eccentricity and a heightened sense of preparedness.

11. TUK-TUK THAILAND

**London Heathrow Airport –
Suvarnabhumi International Airport, Bangkok, Thailand**

Flight time: Eleven hours and thirty five minutes

19 May 2012

Good morning, Bangkok!

As the sun rises over the Chao Phraya River, casting a golden glow on the mist-covered waters, the bustling commotion of Bangkok comes to life. Commuter boats scurry across the river, their lights twinkling in the dawn, while the distant hooting of horns adds an eerie charm to the scene. The river, with its multitude of boats and watercraft criss-crossing its expanse, presents a mesmerising spectacle of chaos and activity, unlike any other river I've encountered in my travels.

After retrieving my toothbrush from the safe, I freshen up for the day ahead, excited to explore the vibrant city of Bangkok. My first meeting is with John, a representative from a prominent donor organisation. His trendy attire, consisting of a loose shirt, jeans, and retro-style glasses, exudes a sense of urban cool more commonly associated with Hackney, London than downtown Bangkok. His warm smile puts me at ease as we sit down to discuss the agenda for a 'strengthening democracy' program aimed at politicians and civil servants from Southeast Asian countries. We recognise the challenges of catering to diverse political systems and varying levels of administrative maturity, but we are determined to give it our best effort.

The morning progresses smoothly, buoyed by John's infectious energy that seems to permeate the room. Just as we are about to conclude our meeting, a politician raises concerns about his country's parliament lacking oversight on government finances, still relying on outdated financial reports from the 1970s. While I am tempted to comment on the government's potential aversion to being held accountable, I choose to bite my lip and focus on discussing viable options for addressing the backlog of financial reports instead.

As evening descends upon the city, I feel the weariness settling in, a combination of the relentless heat, lingering jet lag, and mental exertion throughout the day. I am invited onto a river boat, where I engage in light conversation while attempting to balance a glass of wine precariously in one hand and a plate full of food in the other. The gentle rocking of the boat adds an element of challenge to my juggling act. Just as we pass by the majestic Grand Palace, an architectural gem steeped in history, my balance falters, and the glass of wine escapes my grasp, showering my rather expensive cocktail dress with its contents. The mishap elicits laughter from the other guests, and I join in, choosing to embrace the unexpected moment with a sense of humour.

As the boat continues its journey along the river, the illuminated landmarks of Bangkok creating a magical backdrop, I take a moment to appreciate the city's lively energy and the unique experiences it offers. Despite the occasional mishap, each day in Bangkok is filled with memorable encounters and cultural discoveries that make it a truly enchanting destination.

20 May 2012

Yesterday's workshops ran smoothly, and now it's time for some personal exploration. Jill, a former intern I had the pleasure of working with in London, is currently in Bangkok for her PhD, and we're excited to embark on a city adventure together – a journey through one of the most vibrant and cacophonous cities in the world.

Our whirlwind tour begins and ends with the Chao Phraya River, lovingly referred to as 'the river of kings' and considered the heartline of the city. We hop on and off commuter boats that traverse the river, making stops at various landmarks. First, we visit Wat Pho, home to the magnificent forty-five-metre-long reclining gold Buddha, a sight that leaves us in awe. Next, we explore Wat Suthat, one of the oldest and most stunning Buddhist temples, with its intricate architecture and serene atmosphere.

Before indulging in a late lunch, we make a final stop at the Grand Palace, an impressive complex established in 1782, housing royal halls and a handful of government offices. The opulence and historical significance of the place leave us mesmerised.

As the day draws to a close, we find ourselves at the Sky Bar, perched 820 feet above the ground. This rooftop bar is known for its magnificent views and is one of the highest in the world. Stepping out of the elevator, we are greeted by a mix of impeccably dressed individuals mingling with backpackers. We settle in with mojitos in hand, gazing out towards the river, where hundreds of boats adorned with fairy lights traverse the water. Jill attracts some unwanted attention from a tourist, but we don't let it dampen our spirits. I could spend the entire night here, soaking up the mesmerising views, but the clock strikes 3am, reminding us that it's time to navigate the quiet streets of Bangkok back to the hotel.

Reflecting on the day's adventures, I can't help but feel grateful for the opportunity to explore this bustling city with a dear friend. Bangkok has left an indelible impression on us, with its energy, rich cultural heritage, and sights.

21 May 2012
Waking up with a slight hangover from last night's indulgence in mojitos, I find Jill waiting in the hotel lobby, eager to embark on another adventurous day. It's my final day in Bangkok before I head off to Koh Samet Island, a tropical retreat located 220 kilometres away.

We transition from the refreshing breeze of the river to the stifling humidity of downtown Bangkok as we hop onto a tuk-tuk, an automated rickshaw, that speeds through narrow streets at a breakneck pace. Our pleas for the driver to slow down fall on deaf ears, and we find ourselves tossed from side to side, clinging onto each other for dear life. Despite the recklessness, there's an exhilarating thrill to the ride, and we can't help but giggle, perhaps out of a mix of fear and excitement, as we make our way to Chatuchak market.

Chatuchak market sprawls for miles, offering a vast array of goods and wares. From pets to street food, religious icons to padded bras and jewellery, the market seems to have everything imaginable. The vibrant atmosphere and bustling energy captivate me, and I immerse myself in the experience. Jill and I take shopping to new heights as we dive into piles of colourful clothes and sift through souvenirs. Time flies, and as the day draws to a close, we share a warm hug, parting ways as we each make our way to the sky train.

With memories of the bustling market and the thrill of the tuk-tuk ride still fresh in my mind, I bid farewell to Bangkok, excited for the next leg of my journey to Koh Samet Island. The city has offered me an unforgettable experience, blending cultural richness, culinary delights, and a touch of daring adventure.

22 May 2012

The driver arrives in the foyer an hour late, looking flustered and sweaty from navigating the rush-hour traffic. Just like any other Monday morning, the roads are choked with vehicles and we slowly make our way through the congested streets, finally reaching the highway as we exit the city suburbs.

The journey to Ban Phe port takes three long, uncomfortable hours. The plastic seating sticks to my bare legs, creating a suction effect with every movement, resulting in loud smacking noises. I let out a sigh of relief when we finally arrive at the port.

As I enter the reception area, a tranquil sea of green emerald stretches out before my eyes. A speedboat is moored at the jetty, and the captain patiently waits to transport me to the island. I'm taken aback by the extensive documentation required, and I'm amused by the array of perfume and scented soap options for the villa, ranging from orange blossom and lilac to vanilla, jasmine, and rose. However, when it comes to choosing a beverage, my decision is an easy one; it will always be champagne!

As the boat anchors off the shore, a welcome party emerges to greet me just as I clumsily step off the speedboat, causing a splash in the water. They hide their laughter behind colourful umbrellas, providing shade from the sun. True to their word, the villa is adorned with the fragrance of orange blossom, rose petals delicately scattered on the bed, and a bottle of chilled champagne awaits. Individually scented soaps are neatly arranged in the rain shower, adding a touch of luxury.

Today being my birthday, I couldn't think of a better way to spend it than basking in the sun in this paradise. I make my way down to the white, sandy beach, where the sparkling green water with hints of aquamarine entices me. I spend the afternoon swimming and enjoying the sun, occasionally returning to my sun lounger to respond to birthday messages from friends and indulge in high tea on the beach. Unbeknownst to them, I relish the double portions of tea, cucumber sandwiches, and sponge cake.

As the evening arrives, I witness a spectacular sunset, but my attention is drawn eastward, where a wild thunderstorm brews on the mainland, gradually making its way toward our island. Seated alone on a raised terrace, I find solace in reading a book while the soothing sound of waves breaking over the rocks below accompanies me. Inside the bar, I observe the few other guests, mostly couples and middle-aged Western European men accompanied by young Thai women. Engaging in people-watching, I witness an older man struggling to communicate with a young Thai woman beyond superficial conversations about the weather, besotted newlyweds lost in each other's gaze, and a family attempting to pacify their spoiled children, who complain about their preference for being in Bali.

The thunderstorm finally reaches our island, unleashing its fury with howling winds and palm trees swaying and cowering in response. It feels as if the gusts of wind might tear the roof off. Suddenly, a loud knock interrupts the chaos. Initially hesitant due to the stories I've heard about secluded islands, I cautiously open the door, only to be relieved to find someone holding a large sponge birthday cake. I slice myself a piece, pour another glass of champagne, and engage in a conversation with Mr. Lizard, whom I affectionately name 'Jo'. He proves to be an excellent listener and will become my trusted companion over the next few days. However, I make it clear that the bed is off-limits for Jo.

25 May 2012
Oh, the boredom and loneliness on this island have become unbearable. I can't bear another night of solitude

or another session of people-watching. The only solace I find is in the presence of Mr. Lizard, my listening friend 'Jo', and the *Bangkok Post* to keep me company.

The front-page news catches my attention with the headline, "Lady Gaga hits cabaret amid Rolex tweet fury".[5] Apparently, Lady Gaga spent her first night in Bangkok watching a transvestite cabaret show, while the Thai community was outraged by her purchase of a counterfeit Rolex watch, following her tweet: "I just landed in Bangkok baby and am ready for 500k screaming monsters. I want to get lost in a lady market and buy a fake Rolex." The tweet was seen as offensive and insulting by the Thai people. Alongside this article, there is a more serious piece of news about the former transport permanent secretary, who was found to have acquired unusual wealth of 18 million baht after his house was burgled. When approached for comment, he chose to remain silent.[6]

To fill the remaining time, I treat myself to a sublime massage. From head to toe, the therapist uses sweet-scented oils rich in vitamins and antioxidants. The long, flowing movements of their hands across my body work wonders, untangling my knotted muscles and alleviating my stress. Not that I should have any stress while being in this paradise for so long.

Thankfully, it's time to depart. While paradise is beautiful, it would be even more enjoyable with the company of others. Tomorrow, I set off for Beijing. Farewell, Mr. Lizard, Jo!

---

5   "*Lady Gaga hits cabaret amid Rolex tweet fury*", Bangkok Post, 25 May 2012
6   "*NACC finds against Supoj*", Bangkok Post, 25 May 2012

12. BEDBUG CHINA

Suvarnabhumi International Airport, Bangkok, Thailand –
Beijing Capital International Airport, China

Flight time: Five hours

26 May 2012

It's 5am and we're racing along one of the six four-lane ring roads encircling Beijing. Despite having been here twice before, this part of the city is unfamiliar to me.

As we pass the Bird's Nest, the iconic stadium that hosted the 2008 Olympics, I realise we're heading west of the city. Upon arriving at the hotel, I receive no assistance from the unhelpful concierge, and the receptionist makes me wait while she engages in gossip, showing no intention of using even a smidgen of her English skills. I'll be on my own in this less-than-welcoming place for a couple of days until my colleagues arrive.

In my previous visits to Beijing, I stayed in a central location near the business quarter and iconic landmarks like Tiananmen Square and the Forbidden City, also known as the Imperial Palace. It has been two decades since the student-led pro-democracy protests in Tiananmen

Square, which were met with severe force from the Chinese authorities, resulting in casualties and arrests. The square, now crowded with tourists, stands adjacent to the Forbidden City – a sprawling complex that served as the home of twenty-four emperors over a span of 492 years and is now filled with wooden structures and cultural museums.

My first visit to Beijing was in 2001 when Aidan was working. While he attended conferences, I had the chance to climb the Great Wall of China and zip down on a toboggan at thrilling speeds. I also remember getting slightly tipsy and accidentally ordering twenty-four bottles of milk online from a well-known supermarket upon our return home. Online grocery shopping was still relatively new at the time, so we found it amusing to order supplies from Beijing. Fortunately, they understood our mistake and took the milk back.

Now, I find myself staying in the Shijingshan district, far from the city centre. The choice of this hotel by the organisers seems absurd. After my arrival, I decide to take a walk and explore the area. However, when I ask the concierge for recommendations, he dismisses the neighbourhood, saying, "There is nothing to see here. Head east for a shopping centre or west for a sculpture park – take your pick."

Undeterred by the discouraging advice, I set off towards the west. The air is thick with humidity and pollution, making it difficult to see the sun. Most of the journey, which spans around two miles, is spent walking alongside a busy road, resembling the M1 motorway. Just as I'm about to give up on finding the sculpture park, I spot some trees and grass ahead. Thank goodness, it's the park – a welcome oasis of shady tranquillity.

Scattered throughout the park are forty sculptures donated by various countries to commemorate the 2008 Olympics. The artworks range from contemporary pieces to ornate carvings, reliefs, and modern representations of different cultures. It's a delightful experience to connect with these diverse sculptural creations. This visit serves as a reminder not to always heed naff advice – I have indeed discovered something worthwhile to see around here!

This evening, I find myself alone in the dreary hotel restaurant. The unappetising food has been sitting under an infrared food warmer for far too long. Chicken feet and congealed chicken blood certainly aren't my preferred choices. Hungry, I attempt to order pasta through room service, but to no avail. However, I manage to successfully order a glass of wine, though I'll have to go down to the restaurant to collect it. It's a new take on room service. Frustrated with ordering a meal, I venture out to a local shop in search of a bar of recognisable chocolate.

My attempts to access the internet are futile, likely due to government censorship, so I settle for watching state-run TV. There seem to be only three channels to choose from: modern soaps, historical and cultural programming featuring stories about the Ming dynasty, and the Communist Party propaganda channel, which conveys the government's desired narrative. I opt for the modern soaps, despite not understanding the dialogue. The actors' body language reveals a plot of family dysfunction and misery. Having been away from home for two weeks, I'm overcome with homesickness. I call

Aidan and break into tears, expressing my despair at being alone in this dreadful hotel.

27 May 2012
Oh no, waking up to discover bedbug bites on my leg is a nightmare! The bites resemble large red footballs and are incredibly itchy. I'm amazed that I'm still alive, given the blood spatters on the bed sheets. Quickly, I throw on a pair of trousers to conceal the bites and head down to breakfast. The piles of decaying chicken feet under the heat lamps are still here, ruining my appetite. Thankfully, some colleagues have arrived this morning, and I feel relieved to be in the company of friends again. We discuss the hotel's pitfalls over breakfast, and many are already considering moving out upon learning about the lack of internet access and seeing my troublesome bites.

The day flies by with work meetings, and soon it's time for us to be taken to the Huguang Guild Hall, one of the best wooden theatres in Beijing and once the centre of political and social life, to witness a performance of 'the rainbow skirt and feathered coat dance'. This romantic and enchanting dance has its origins dating back thousands of years to the Tang dynasty. As the performance begins, the stage comes alive with vibrant colours and intricate costumes. The dancers move with such grace and speed, twirling around like whirling dervishes. Watching them spin so rapidly makes me feel a bit dizzy.

In recent days, I've been surviving on the occasional chocolate bar, and now I'm craving something familiar and spicy to eat. Anna, who was with me in South Korea, suggests that I try the spicy noodles. They look delectable in the menu picture, but when they arrive, they resemble a heap of thick, gelatinous noodles that remind me of jellied eels. My appetite takes a nosedive. However, I don't want to be rude, so I decide to give them a try. Oh my goodness, as soon as they touch my mouth, it feels like I've eaten an entire Scotch bonnet chilli, and my lips quickly start to swell and feel hot. Anna looks horrified and rushes to fill a glass of water to help soothe the effects. I settle for the green salad for the remainder of the evening, hoping for some relief.

28 May 2012

I find myself in an unexpected situation as I present at a conference near the Bird's Nest stadium. The auditorium is supposedly filled with one thousand delegates interested in accounting, or so I was led to believe. However, it becomes apparent that the majority of attendees are members of the public who were chosen and paid by the communist party to be there. It's a rather peculiar setup, but at least it helps to maintain appearances in front of international delegates. The whole situation feels quite surreal.

As if things couldn't get any stranger, when I return to my hotel room, I discover that the door was left partially open. With caution, I enter and immediately notice that someone has been rummaging through my belongings. Certain items have been moved from their original positions. This discovery leaves me feeling uneasy, especially considering China's reputation as a surveillance state. I highly doubt it was the cleaners, as they would have closed the door and respected my personal space. It seems my privacy has been ominously violated, and now I find myself concerned about falling asleep, fearing a potential anonymous visit. To enhance my sense of security, I place a chair behind the door and stay awake for as long as I can, attentively listening to every sound. The telephone rings multiple times, with the same person on the other end enquiring about my departure time in broken English. It becomes increasingly apparent that my every move is being monitored.

29 May 2012

After a night filled with restlessness and anxiety, I feel a sense of relief as I finally embark on my journey to the airport. It's early in the morning and exhaustion weighs heavily on me, exacerbated by the discomfort caused by the persistent bug bites. Nevertheless, the knowledge that I'm headed home brings solace.

Travelling can be a remarkable experience, but no matter where I find myself in the world, there's always a special feeling when I step onto the plane that will take me back home. The anticipation of returning, unlocking my front door, and setting down my bags is unparalleled bliss. This time, the desire to reach home as quickly as possible is overwhelming.

13. MALDIVES – WIGS AWAY!

**London Heathrow Airport –
Zayed International Airport, Abu Dhabi – Velana International
Airport, Malé, Maldives**

Flight Time: Twelve hours and twenty-five minutes

9 October 2012

As I settle into my seat on the plane, preparing for the second leg of my journey to Malé, I find myself unexpectedly approached by an MP and his policy adviser from Bangladesh, who insist on taking a photo with me. My attention is immediately drawn to the MP's poorly fitted toupee, which seems to have a mind of its own, resembling either a flattened hedgehog or a questionable rug from a discount store. Despite my efforts to maintain political correctness, I find it difficult to avert my gaze from the peculiar sight.

Only a few weeks ago, an email landed in my inbox, inviting me to participate in a capacity-building program in the Maldives organised by an international development organisation. The prospect of visiting paradise is irresistible, prompting me to respond with enthusiasm, accompanied by a thumbs-up emoji.

Upon landing, our group is swiftly escorted to the VIP suite, where every detail is meticulously taken care of. We breeze through security and board a speedboat that transports us to one of the Maldives' stunning islands, evoking memories of the exhilarating speedboat scenes from the Bond film, *Live and Let Die*, where Roger Moore fearlessly navigates treacherous waters.

The sea proves to be tumultuous, causing our boat to jolt and toss us around like rubber ducks in a bathtub. Waves crash, winds gust, and sprays of water sting our faces. In an effort to stabilise the boat, we are asked to change seats, providing an opportunity for me to discreetly observe my fellow passengers. I can't help but notice that the MP's toupee, or rather, the rug, has completed a full rotation on his head. It takes all my self-control to suppress my laughter as I avoid making eye contact. I'm amazed that the wig hasn't been swept away by the wind. With a mix of relief and amusement, we disembark from the boat, shaken but not stirred by the turbulent ride.

10 October 2012

The day begins with an early start as workshop activities fill up most of my schedule. Stepping out of the villa, I'm immediately captivated by the magnificent view of the crystal-blue sea stretching into the distance and the pristine white sand beach. It truly feels like paradise. As I make my way to the venue, I can't help but notice the curious gazes of holidaymakers lounging on the beach, seemingly surprised to see someone dressed in a business suit, carrying a briefcase amidst their relaxed vacation ambiance.

The meeting gets off to a promising start and I'm pleased to see a good representation of women in attendance. However, I can't help but notice that the majority of female participants are from the international development agency and committee clerks supporting the event, rather than the politicians and experts themselves. Throughout the discussions, I occasionally sense a lack of active listening when I speak, particularly from some male participants who seem more attentive to their male colleagues during their presentations. I can only speculate on the reasons behind this disparity, but it raises important questions about gender dynamics in such professional settings.

During dinner, I engage in a heartfelt conversation with one of the participants who recently experienced the loss of their partner. It becomes evident that this is their first time being away from home since the tragedy. As we walk back from the evening meal, I witness his tears and overwhelming emotions. In an attempt to provide comfort, I suggest that he calls home and say hello to his children, knowing that such moments of connection can bring solace and warmth during difficult times. It saddens me to see individuals grappling with personal turmoil while having to travel far distances for work obligations.

Reflecting on the day, it becomes apparent that even in the midst of professional engagements and workshops, personal struggles and human connections remain significant aspects of our lives. The juxtaposition of the idyllic setting with personal challenges reminds me of the complexities we all carry within us.

11 October 2012

Today, we embark on a journey to Malé, the bustling capital city of the Maldives. The cityscape is dominated by high-rise buildings, and every available space seems to be utilised. Interestingly, cats and dogs are not allowed in Malé, adding to the unique character of the city. The Maldives, idyllically located in the Indian Ocean, heavily relies on tourism for its economy. However, the islands face the daunting challenges of economic poverty and the threat of climate change. With rising sea levels and increasing temperatures, the low-lying nature of the Maldives puts it at great risk. A significant portion of the islands are just a metre above sea level and annual flooding is a recurring issue, highlighting the urgency of addressing climate change. In fact, the government once held a Cabinet meeting with ministers donning scuba gear to raise awareness about the devastating impact of climate change on small island nations.

We make our way through the narrow streets to reach the People's Majlis, the Parliament of the Maldives. Tucked away in a side street, the small but significant building holds historical importance. It was here that President Mohamed Nasheed resigned in February following three weeks of opposition-led protests ending in police mutiny. These protests were part of the broader Arab Spring movement, which swept across many Arab countries, driven by frustrations over corruption and economic stagnation.

Boarding a boat, we set sail to Hulhumalé, a man-made island created to alleviate the congestion of Malé. The island, still in its developmental stages, appears flat and barren. Construction is ongoing, and the landscape is marred by litter and plastic bottles, painting a stark contrast to the picturesque image of the Maldives. The predominant industry on Hulhumalé is a tuna fish processing plant, where we witness the laborious process of handling and canning these large fish.

A government official shares their vision of relocating around 300,000 Sunni Muslims to Hulhumalé in the next two decades, with aspirations of transforming it into the 'Mega City of Asia'. Another representative likens it to the 'Hong Kong of the Maldives'. These ambitious plans reflect the government's efforts to address population growth and urban development in the face of limited space and resources.

In the evening, we attend a lively beach party on a nearby island adorned with palm-fringed shores. The attendees are dressed to impress, with Bhutanese guests showcasing their stunning traditional attire. The men don the knee-length robe called *Gho*, reminiscent of a kimono, while the women wear the elegant ankle-length dress known as *Kira*. This attire represents the official dress code for formal events in Bhutan. Meanwhile, the men from Afghanistan exude style with their neatly trimmed beards, turban headwear, and vibrantly patterned coats. Amidst this vibrant gathering, I can't help but feel slightly underdressed in my jeans and T-shirt. Nevertheless, the atmosphere is electric, and the festivities kick off in full swing.

Tables adorned with delectable food line the beach, illuminated by torches casting a warm glow. The cool sand beneath our feet adds to the sensory experience as musicians and dancers captivate us with their rhythmic performances accompanied by resounding drums. In the flickering light of the beach fires, a colleague from Afghanistan invites me to dance, and with bare feet in the sand, we join in the spirited celebration. This enchanting evening exemplifies the joy and camaraderie that can arise from connecting with people from diverse cultural backgrounds. It serves as a reminder that despite the challenges we face, embracing and appreciating our differences can create moments of unity and happiness. This memorable night will be cherished for years to come.

12 October 2012

It is disheartening to learn about the allegations of sexual harassment that have emerged during the workshops. The atmosphere in the camp is filled with unease, and it becomes evident that something untoward has been happening when I am directed to a dining table exclusively for women.

To my surprise, the person whom I had empathised with on the first evening turns out to be the alleged perpetrator. Since arriving on the island, he has been engaging in sexual harassment, pressuring young women to sleep with him or to find others who would comply. It is truly saddening to witness such behaviour, where women are subjected to harassment and feel afraid to report it to their powerful male superiors for fear of job loss or retaliation. Unfortunately, instances of sexual harassment are not uncommon in events like these, and they persist worldwide.

What is crucial is the establishment of clear ground rules and expectations from the outset of such events to prevent such situations from occurring. Institutions responsible for funding international programs should have robust policies in place to address and combat sexual harassment, taking a proactive approach to eradicating this behaviour. It is vital to create a safe and inclusive environment where everyone feels respected and protected from any form of harassment or discrimination.

13 October 2012

It is quite a surprising turn of events to witness the accused individual of sexual harassment walking outside the young women's villa in an outfit that includes burgundy speedos and an elaborately embroidered silk dressing gown. Rather than feeling upset or threatened, the young women find it comical and burst into laughter. The absurdity of his attire seems to have stripped him of any remaining authority, allowing the women to find some form of revenge through humour. It's good to see them finding joy and reclaiming their power in this situation.

14. THE BRUSSELS HOPS

St Pancras Station, London – Brussels Midi Station, Belgium

Train time: One hour and fifty-three minutes

14 January 2013

This won't be my first trip to Brussels and it certainly won't be my last. I've grown accustomed to making this journey at least three times a year. However, with each visit, it becomes increasingly tiresome, not because Brussels itself is boring, but because I have to drag myself out of bed at an ungodly hour to catch the 6am train from St Pancras Station, just to arrive at the Accountancy Association for a 10am start. The routine remains the same: a day spent in a meeting room with colleagues from across Europe, followed by a rushed two-hour lunch, and a frantic race back to Midi Station for the 6pm train.

Even at this early hour, the train is packed with commuters heading to the European Commission or embarking on corporate business trips. The enticing aroma of hot, buttered croissants, sizzling bacon, and freshly brewed coffee fills the train carriages as we enter the Channel Tunnel. The delightful scents help keep me awake as I browse through the agenda papers.

Initially, I thought I would need to brush up on my French language skills for these trips, but as all the business is conducted in English, I grew lazy and abandoned my French language course. Perhaps if they rotated languages for each meeting, it would encourage everyone to make more of an effort. Upon arrival in Brussels, I find myself standing in the lengthy taxi queue that winds around the station entrance, only to rapidly diminish as taxis pull up. Today, at least, I won't be rushing back for the 6pm train since I'll be presenting in the European Parliament. However, the brief provided by Maria, who oversees the Brussels office, is quite vague, which concerns me from the outset as it may lead to my presentation being off message.

Every time I visit Brussels, I never seem to have enough time to truly explore the city. Any shopping I manage to do is usually last-minute at Midi Station, where I can never seem to find that elusive Tintin T-shirt, paying homage to the fictional Belgian reporter and adventurer. However, as I gaze out of the taxi window, I'm reminded of the city's beauty. In addition to the towering glass buildings of the European Commission, there are stunning architectural gems. The most striking features on the old buildings are the exquisite Art Nouveau doors, each a work of art in itself, boasting a variety of shapes, sizes, colours, and intricate floor tiles. These distinctive elements set Brussels apart from any other city in the world.

Upon arriving at the European Parliament, I'm greeted by a flurry of MEPs and policy advisors rushing about, clutching stacks of briefing papers. Young women are dressed in fashionable attire, donning high heels and short skirts. I can still recall the time in the 1990s when trouser suits were in vogue. I boldly broke the dress code of a former employer by being the first female to wear a trouser suit, a fashion choice I would embrace throughout the rest of my career.

We are directed to one of the largest committee rooms, where I meet the other speakers, all MEPs. The panel session commences, and to my dismay, the topic is not what I had anticipated. The MEPs seated next to me had engaged in an earlier meeting discussing grant aid and continuously refer to their prior discussion. Unfortunately, I was unaware of their previous conversations or held any specific views on grant aid, making it challenging for me to contribute positively to this debate. What I had prepared to discuss is only tangentially related to the subject and may come across as out of left field. It's one of those moments when I wish the ground would open up and swallow me whole. Strangely, Maria sits calmly in the audience, as if everything is going according to plan. However, judging by the body language of the other panellists, especially their piercing stares fixed upon me, I know I have missed the mark by a wide margin.

The feeling of relief washes over me as the session finally concludes. The only silver lining is that we receive decent press coverage. Exhausted, I board the train for my journey home, indulging in copious amounts of Belgian chocolates and sparkling wine. Instantly, I start to feel much better.

15. KATHMANDU – MY NOT-SO SHANGRI-LA EXPERIENCE

London Heathrow Airport – Hamed International Airport, Doha, Qatar – Tribhuvan International Airport, Kathmandu, Nepal

Flight time: Thirteen hours

6 March 2013

It's 1am when our plane lands at Hamed International Airport in Qatar, the first leg of my journey before heading to Nepal. This trip is focused on providing support to individuals responsible for managing government finances. Upon arrival, the bustling business lounge quickly empties, leaving only a few of us waiting for the morning flights. The loud vacuum wielded by the cleaners prevent me from falling asleep, so I take the opportunity to respond to work emails until our 5am departure. Emailing colleagues throughout the night has become a bad habit of mine, as airports are never particularly interesting. As I press the send button on my bulk message and close the computer, the flight board suddenly springs to life.

Just as I settle in for the final leg of the journey, I'm interrupted by a Nepalese man who seems eager to discuss his country's politics. I had done some research prior to the trip and had a basic understanding of Nepal's history, including its turbulent civil war and relatively new democracy. However, what I didn't anticipate was landing in the midst of a three-day public strike. According to my newfound Nepalese acquaintance, I may encounter difficulties leaving the airport and reaching the hotel. He tells me, "I'm walking home from the airport to avoid the risk of my family driving a car that could be targeted by demonstrators." This revelation leaves me deeply concerned.

We touch down, and I find myself standing alone in the airport arrivals lounge, the sole Westerner and female in sight. Military police armed with automatic weapons, armoured vehicles, and tanks surround the airport perimeter. There's no sign of my driver, and my mobile phone has run out of battery. At this point, most sensible individuals would turn around and catch the first flight out of here. However, perhaps stupidly, I throw caution to the wind.

I spot a small operating taxi kiosk and, seeing no other signs of activity, I cautiously approach to order a taxi. It appears legitimate, so I decide to take the risk. A battered, three-wheeled white van resembling those used for transporting olives in the southern Mediterranean pulls up. My suitcase is unceremoniously tossed into the van, and I take my seat behind the driver, bracing myself for what lies ahead.

We speed past the ring of military police, exiting the airport and onto the wide dirt tracks that serve as roads in Kathmandu. There are no other vehicles in sight as we bounce over rubble and stones, hurtling along at breakneck speed. The driver, a burly man, perspires profusely and appears visibly anxious. My worry intensifies as I begin to contemplate various worst-case scenarios. What if I vanish without a trace? What if I'm wrongfully arrested and detained? Who would even know that I took a taxi, and how could they retrace my steps?

Crowds of people line the roadsides, amidst a sea of rubbish ranging from plastic bottles to decaying fruit. The driver honks his horn, warning pedestrians not to step onto the road as we pass by. The vehicle lurches and jerks as I catch sight of army personnel and tanks up ahead. A military officer approaches our van, capturing a video of us amidst the throngs of people on the streets.

The driver is drenched in sweat, appearing on the verge of a heart attack. My confidence in our safe passage dwindles. Nevertheless, he persists, and I find myself rigid with fear as I shout at him to steer clear of the crowd. He frantically beeps the horn, causing people to scatter, and we swerve to the right. It's a close call, but we manage to continue our journey through narrow streets, littered with even more rubbish, and cross a bridge over an open sewer. The stench is putrid and sour. This is far from the Shangri-La I had hoped for!

Finally, we make a left turn into the hotel grounds, and as the tension releases from my shoulder muscles, the anxiety starts to dissipate. We come to an abrupt stop. In a gesture of gratitude, I hand the driver a generous tip, likely equivalent to a month's wages, to express my sincere appreciation for his efforts.

7 March 2012

As I sit up in bed this morning, I find myself engrossed in the local newspaper, the *Himalayan*. The front-page story captures my attention, recounting the tale of a snake that unexpectedly takes up residence in a family's home, initially welcomed as a harbinger of good fortune. However, the snake's untimely demise leaves the family devastated, fearing a streak of misfortune. It's a refreshing diversion from the usual news of economic austerity back in the UK. Despite their contrasting subjects, both headlines share the common theme of being equally disheartening.

By midday, I am scheduled to meet with colleagues coming from diverse regions across South Asia and Australasia, alongside officials from the Nepalese government. The purpose of our gathering is to discuss public financial management reform, intending to facilitate the exchange of practices and experiences. However, it has become increasingly apparent that the discussion has turned into a one-way street, with government officials reluctantly participating. Thomas, who chairs the session, struggles to engage them, and is met with prolonged silences in response to his questions. Frustration creases his brow, reflecting his exasperation and anger. Perhaps the lack of a stable government, coupled with the ongoing public strike, preoccupies their minds, overshadowing their focus on government accounting. The meeting becomes a trying and arduous affair.

Later, as I walk past the public bar, I notice my male colleagues gathered in a tight circle, undoubtedly engrossed in discussions about the day's events and conducting informal business dealings outside of the formal meetings. This occurrence is not uncommon during my travels. It is within these informal networks that pivotal decisions are often made and the following day's agenda is finalised. I don't believe they consciously exclude me or other female colleagues from these discussions; rather, it seems they are unaware of how uncomfortable it can be for women to approach and integrate into their conversations. I have attempted to convey this point on numerous occasions.

8 March 2013

Alex, a colleague and I eagerly await the departure of the Buddha airplane for our much-anticipated sightseeing trip to the majestic Himalayan mountain range and Mount Everest. We arrived at the airport bright and early at 7am, but due to a lingering morning mist blanketing the city, flights have been grounded for hours. There is an air of uncertainty, and the possibility of losing our $200 refund looms. However, just when hope seems to fade, the mist miraculously lifts and our flight number appears on the bustling flight board. Joyously, we realise that we are finally on the move.

The aircraft is small, accommodating a maximum of forty passengers. As we brace ourselves, it rumbles and trembles down the runway, lifting off into the eastern skies. Nepal airlines may have a dubious safety rating, and the memory of a recent tragic plane crash lingers, but we choose to embrace the excitement and take the risk.

As we soar through the air, I glance down at Kathmandu, a sprawling city stretching across the length of the valley. Then, turning my gaze to the right, my eyes are met with an incredible sight – the Himalayas. Their grandeur and immensity are awe-inspiring, especially against the backdrop of a clear, cloudless sky.

The pilot directs our attention: "Look to the left, and you will spot Gosaithan, towering at 26,290 feet. Just to its right is Dorje Lhakpa, a majestic peak at 22,000 feet above sea level." He enlightens us about the spiritual significance attached to each mountain and shares stories of the first climbers to conquer them. As we approach Mount Everest, reaching a staggering height of 29,029 feet, we are blessed with an unobstructed view. Its limestone summit commands attention, towering above all other peaks, with icy trails cascading down its slopes. It is an unforgettable sight, bringing us within reach of this ancient mountain, the highest on Earth. The pilot reveals that Everest's height is ever-changing due to the shifting tectonic plates, causing solid rock slabs to fracture – a geological phenomenon known as a 'tectonic smash-up'. Remarkably, this process propels India to creep northwards by about two inches every year.

Inviting me to the cockpit for a photograph, the pilot extends a kind gesture, though I can't help but feel a touch of trepidation. After all, I would prefer he focus on flying the plane, especially with the fierce headwinds surrounding Everest. Alex, who had dozed off during most of the mountain viewing, wakes up just as we touch down. We are soon joined by Emily, another colleague, as we disembark the aircraft.

With an insatiable thirst for adventure, we venture to the royal palace, Hanuman Dhoka, to explore its peculiar museum of royal artifacts and seek solace from the scorching heat within the palace rooms and courtyards. Our journey continues as we make our way to Swayambhunath, a magnificent Buddhist temple and

UNESCO World Heritage site perched atop a lofty hill. As we arrive, mischievous baboons playfully surround us while we navigate the stone steps, carefully stepping over slumbering dogs seeking shade. The gleaming white stupa and gilded spire, adorned with Buddha's all-seeing eyes, shimmer in the sunlight. The scent of incense and flickering butter lamps fill the air, and at every turn ornate carvings, religious symbols, and whimsical curiosities captivate our senses.

No visit would be complete without indulging in the purchase of pashmina shawls. Pashmina, the finest type of cashmere wool derived from goats, is often woven or hand-spun in India and Nepal. These beautiful textiles are readily available and affordable in Nepal, although Western tourists are now encountering higher prices. Emily and I find ourselves irresistibly drawn to the softness and allure of these shawls, draping them over our shoulders in delight. The array of choices overwhelms us as the shopkeeper seizes the opportunity to make a sale. His counter is adorned with pashminas of every colour, pattern, and stitching, and at the slightest hint of interest, he feverishly punches numbers into his calculator, eager to present us with his best price.

Time slips away as we spend over an hour in the shop, much to Alex's growing impatience. Unlike us, he swiftly made his scarf selection within five minutes. We heed the sage advice to ensure our scarves are crafted from 100% cashmere, as anything less would be a grave fashion faux pas.

16. AWESOME CANADA

**London Gatwick Airport –
Ottawa International Airport, Canada –
Toronto Pearson Airport, Canada**

Flight time: Nine hours and forty-eight minutes and one hour and fifteen minutes

7 June 2013

I can't help but roll my eyes as the waitress enthusiastically declares that ordering a cocktail is 'awesome'. Really? It's just a cocktail, not a life-altering experience that would shake the foundations of the world. If I were to win the Nobel Peace Prize, that would be 'awesome'. But the word lingers in my mind, and before I know it, I catch myself using it repeatedly.

Here I am in Ottawa, the capital city of Canada. Despite it being May, there's a biting wind in the air and patches of snow stubbornly cling to shaded areas. Earlier today, I was frantically running from one government building to another, interviewing civil servants as part of my comparative research on how parliament holds the government accountable for public spending.

This city exudes a similar aura to the UK's 'Westminster Village', where a tight-knit community of MPs, journalists, peers, and researchers converge. Canada's parliamentary system is derived from Westminster, consisting of the Crown, the Senate, and the House of Commons. However, a key distinction is that, as a federal state, Canada shares legislative powers between federal and provincial governments. Like the UK, Canada is no stranger to its own parliamentary scandals, including controversies surrounding parliamentarians' expenses. *The Globe and Mail* newspaper has been critical of the sluggish pace of investigations into a senator's travel expenses.[7]

My research efforts are rewarded with a guided tour of the parliamentary estate. I've lost count of the number of parliaments I've toured over the years – they seem to blend together at times. The UK Parliament, the Scottish Parliament, the Welsh Assembly, Stormont, the European Parliament, the Hungarian Parliament, the Maldives Parliament, and now the Canadian Parliament. I wonder what unique aspects this visit will bring.

From the outside, the parliamentary building stands proudly with its formidable Gothic facade, overlooking the expansive Ottawa River. Our guide informs us that the current structure is not the original, as a fire ravaged the building in 1916. As we step inside and wander through the corridors, we encounter two familiar chambers – the House of Commons, adorned in red, and the Senate with its green hues. Continuing our exploration, we are treated to a mix of architectural styles, including Gothic windows,

---

7   *"MP's exit turns pressure on Harper"*, The Globe and Mail, 7 June 2013

majestic doors, towers, iron finials, and ornate crests. While the tour proves to be fascinating, my grumbling stomach reminds me that it's time to join my colleagues for a high-calorie meal while passionately cheering on Team Canada as they face off against the US in ice hockey!

8 June 2013

Ottawa being the administrative hub of the country, it comes as no surprise that it frequently plays host to government finance conferences. During these events, the city is inundated with a swarm of public sector accountants, clad in their signature black and grey suits, resembling a sea of locusts descending upon the city. I received an invitation to attend this conference due to my contributions as a guest writer for a well-known government financial management magazine. Today, I will be presenting on one of my published articles.

However, it wasn't my presentation that turned out to be the highlight of the day. It was the subsequent discussion I found myself caught up in regarding the process of making a small donation in Canadian dollars to a UK charity. Funding non-Canadian charities seemed to lack a clear precedent. Nevertheless, after some deliberation and collaborative effort, we managed to find a satisfactory solution.

9 June 2013

Toronto truly exudes a cosmopolitan vibe that sets it apart from Ottawa. It serves as a melting pot where individuals from various parts of the world and diverse cultures come together to call this city home. The energy in Toronto is palpable, and its vibrant atmosphere is evident in the abundance of modern art galleries and museums that dot the cityscape. Even beneath the surface, Toronto offers a network of subterranean spaces, providing sheltered shopping and arcades for city dwellers during the harsh winter months.

However, my arrival in Toronto coincides with a period of political turmoil. The city's mayor, Rob Ford, has been ousted from office following the release of a video showing him smoking crack cocaine. Additionally, he faces accusations of using city agencies to benefit his family's business. In public appearances, he appears erratic and vehemently denies all allegations. Earlier in the year, he made headlines for voting against granting honours to both the Canadian Olympic Team and Nelson Mandela, later claiming to have pressed the wrong voting button – a controversial move, to say the least.

Adding to the turmoil, Doug Ford, Rob Ford's brother and Vice Chair of the 'Build Toronto' Agency, witnessed the sudden departure of half of the agency's thirteen-member board. According to reports from the *Globe and Mail*, the agency now finds itself in a state of chaos, scrambling to find new board members. However, Doug Ford maintains an optimistic outlook, stating that things are going well.[8]

As I pass by Toronto City Hall, I observe council officers who have been dismissed, their heads bowed and

---

8   *"Ford optimistic about search for agency staff"*, The Globe and Mail, 7 June 2013

tears in their eyes, clutching brown cardboard boxes filled with their office belongings. It is a sombre scene, reflecting the impact of the ongoing political upheaval.

10 June 2013

Today marks my final day, and I'm eager to witness yet another natural wonder of the world – Niagara Falls. Like the 12 million tourists who flock here annually, I'm ready to immerse myself in the awe-inspiring beauty of this magnificent site. Stretching across Ontario and the state of New York, these falls are a sight to behold.

The Canadian side boasts the grandest of the falls, where a staggering 90% of the water plunges into the gorge from a height of 170 feet. As I gaze downwards at the fast-flowing water, a sense of unease washes over me, causing my legs to feel unsteady. Peering into the depths, I catch a glimpse of a cluster of yellow raincoats bobbing up and down on boats, carrying eager tourists attempting to capture the majestic cascade in their photographs. However, the forceful spray from the falls threatens to knock them off their feet, leaving them drenched and bedraggled.

This excursion serves as a fitting conclusion to a bustling week, and it also allows me to cross off yet another item on my ever-growing 'bucket list'. Truly, it's an experience that is 'awesome'!

17 SPRINGBOK IN JOHANNESBURG

**London Heathrow Airport –
O.R. Tambo International Airport, Johannesburg, South Africa**

Flight time: Ten hours and fifty-five minutes

25 August 2013

The last time I set foot in South Africa was a staggering twelve years ago during my visit to Cape Town. Now, as I traverse the bustling streets of Johannesburg en route to our hotel, I can't help but ponder the changes that have taken place in the country since my last visit. The stark disparities that once defined South Africa come flooding back to my mind. Cape Town, in particular, was a city of stark contrasts, where affluent leafy suburbs, outstanding mountains, and picturesque coastal scenes stood in stark contrast to vast stretches of windswept flatlands, dominated by mile after mile of slums known locally as Cape Flats. Although successive governments have endeavoured to address this segregation since the end of apartheid, the imprint of inequality remains visible.

Even in my brief time in Johannesburg, I can already sense the presence of lingering inequalities. The city exudes an edginess that can be attributed to its distinct features. Properties the size of mansions are ensconced behind towering walls and fortified with barbed-wire fences, while armed security guards and their vigilant canines patrol the streets. It's no wonder that Johannesburg has earned a reputation as a 'lawless city'. Yet, amidst these observations, I acknowledge that Johannesburg stands as South Africa's largest city and economic powerhouse. Over the next few days, I am eager to discover the city's resilience and inherent beauty, as well as to meet the locals who take immense pride in their vibrant home.

Upon reaching our hotel in the western part of the city, we encounter an unexpected setback. Much to my dismay, I learn that I won't be able to check in until late afternoon. After a long and tiresome flight, I am left feeling annoyed, fatigued, and somewhat unkempt. It's disheartening to realise that I must endure another two hours before I can finally access a room and freshen up before meeting the team.

26 August 2013

It's shaping up to be a jam-packed day, starting with a team meeting that promises to be more thrilling than the latest episode of a political drama. We'll be rubbing shoulders with politicians and civil servants from various government institutions, all in an effort to brainstorm how we can entice talented individuals into public-sector finance. Oh, the allure! As if that weren't exhilarating enough, we're also determined to tackle the pressing issue of high unemployment and the high vacancy levels in local government finance. It's a battle against the odds, but we unanimously agree that making a career in this field enticing for young people is a step in the right direction. If we don't address these challenges, the government be stuck in a never-ending loop of repeating the same issues year after year.

After a quick lunch that's devoured in record time, we embark on our getaway to the high-security car park. Sarah, my ever-vigilant colleague, becomes our very own secret agent as she scans the surroundings, ensuring no carjackers are lurking nearby. Because, you know, carjacking is just your average daily occurrence here in Johannesburg. Once we're confident the coast is clear, we hop into the car and speed off to give a press interview, where we get to enlighten the world about the importance of strong public finances. Move over, Hollywood action movies – we've got our own adrenaline-fueled adventure right here!

Finally, with a brief respite from the world of finance, I find myself wandering off to a wildlife conservation park. Lions, cheetahs, hyenas, and jackals roam freely within its boundaries, reminding me that life can be both beautiful and terrifying. I prefer to admire these majestic predators from the safety of my locked car, where they loom close to the windows of our trusty Jeep. As one of them saunters over, I suddenly find myself face to face, my eyes locked with its fierce gaze and its jaws that could probably crush an entire watermelon with ease. Even behind the protection of the car's glass, the encounter manages to send a chill down my spine.

As a special treat, I'm granted permission to stroke a one-year-old lion cub. But let's not picture a cute little ball of fur that fits snuggly in the palm of my hand – oh no! This teenager of a cub towers over me, with a powerful build that puts my waistline to shame. With trembling hands, I cautiously step forward alongside the ranger, extending my hand to stroke the cub's rough, wire brush-like coat. I carefully avoid its tail and neck, mindful not to provoke the powerful beast. "I value my hands," I mutter nervously, as if the cub understands the importance of my typing skills. After a few gentle strokes, I retreat to the safety zone, my pulse still racing.

Later in the evening, my colleagues enlighten me with a fascinating titbit: the springbok, a small gazelle I had spotted earlier, holds the prestigious title of South Africa's national animal, and proudly lends its name to the country's rugby team. But here's the twist – the springbok emblem was also once a symbol of oppression during the dark days of apartheid. Who knew that a graceful gazelle could be so entangled in history? Equally intriguing is the fact that a highly spirited alcoholic drink shares the springbok name, and let me tell you, it has a kick that matches the gazelle's leaps and bounds. Crafted from amarula cream and peppermint liquor, this potent concoction surprises your taste buds with a fruity caramel sensation courtesy of the marula fruit. It packs a wallop, but you won't fully grasp the impact until you've downed three shots, slipped off your bar stool, and awakened the next morning with a headache that would rival the power of an elephant's footsteps.

Ah, South Africa – a land of exhilarating challenges, incredible wildlife, and unexpected encounters, where even the drinks pack a delightful punch. Cheers to the adventure!

18. XI'AN, CHINA: DEER GOULASH AND DISAPPEARING DELEGATES

London Heathrow Airport –
Helsinki-Vantaa International Airport, Finland –
Xi'an Xianyang International Airport, China

Flight time: Twelve hours

12 September 2013

Here I am, at the airport, ready to embark on the next leg of my journey to Xi'an – pronounced 'shee ahn' in case you were wondering – located in Central Northwest China. Most people know of this city because of the famed Terracotta Army, which has somehow landed a spot on my never-ending 'bucket list'. I read somewhere that Xi'an holds the prestigious title of being the oldest of China's four ancient capitals, and to top it off, it was even featured in the esteemed *Economist* magazine as one of the thirteen emerging megacities. As usual, I find myself invited to yet another conference, where I get to enlighten academics and business leaders with my profound public finance wisdom. Or at least that's what I like to believe.

This marks my third visit to China, and fingers crossed I won't encounter any covert searches of my personal belongings this time around. The last time it happened, I reported the incident to the Foreign Office and to my surprise, they weren't fazed in the slightest. They casually advised me to give them a heads-up if I ever planned on returning. Of course, being the organised individual that I am, I completely forgot to do so before this trip. So, caution will be my middle name throughout this adventure.

As I wait at Helsinki-Vantaa International Airport, I can't help but notice the throngs of Chinese people eagerly clutching their shopping haul, overflowing with major high-street brands and designer labels. It's a sight to behold, though I must admit, the amount of single-use plastic accompanying their purchases is disheartening. It's clear that their buying power has soared as China's economy continues to thrive. It's like witnessing a shopping extravaganza on steroids.

Settling into the flight, armed with a captivating book, I eagerly dive into its pages, craving an escape from the mundanity of air travel. Alas, my tranquil reading is interrupted by a flight steward offering deer goulash for lunch. For a brief moment, the thought of consuming Bambi crosses my mind, leading to a sudden pang of guilt. Nevertheless, I suppress my emotions and indulge in the unconventional meal.

In a valiant attempt to alleviate the boredom of the flight, I strike up a conversation with a fellow passenger – an affable Austrian gentleman from Innsbruck. Little did I know that beneath his friendly exterior lurked a racist. Yes, he turned out to be anti-everything. He had a bone to pick with Islam, Judaism, Britain, and even the Chinese. Ironically, he had no qualms flying halfway across the globe to conduct business with the very people he openly despised. Oh, the contradictions of humanity! Now, I find myself regretting my decision to engage him in conversation as he spews his divisive opinions, leaving me longing for some peace and quiet amidst the clouds.

Upon our arrival in Xi'an, I'm whisked away to yet another bland and generic business hotel. It's funny how no matter where you are in the world, they all seem to blend into one another – a never-ending parade of indistinguishable accommodations. As I settle into my room, I treat myself to a well-deserved glass of wine and a decadent bar of chocolate. With these companions in hand, I peruse tomorrow's agenda, searching for enticing conference sessions to attend. However, my eyes land upon the first session, which disappointingly lacks any listed speakers. Ah, the perfect excuse to conveniently skip the early morning affair and allow myself a bit more time to recover from the dreaded jet lag. After all, self-care is of utmost importance, isn't it?

13 September 2013

I find myself rudely awakened by an incessant screeching alarm at the ungodly hour of 6am. No recollection of requesting an alarm call prompts me to roll over, desperately seeking solace in a few more moments of blissful slumber. Silence ensues momentarily, teasing me with the prospect of respite, only to be shattered once again by the piercing screech. Now fully awake and seething with fury, I begrudgingly decide to attend the morning session, fuelled by nothing more than sheer curiosity. Grabbing my trusty toothbrush from the safety of the hotel safe, I hastily get ready and prepare myself for the day ahead, before venturing out to the conference alongside a staggering two thousand other delegates. Like a grand procession, we are escorted through the smog-laden streets of Xi'an by a police convoy, making our way to an exhibition centre of gargantuan proportions for the Euro Asia conference.

Despite its promising title, the conference isn't about the fascinating convergence of East and West. No, no! The West apparently stops dead in its tracks at the borders of Ukraine. Instead, what we have here is a gathering of countries that proudly bear the suffix '-stan' in their names – Afghanistan, Kazakhstan, Tajikistan, Uzbekistan, Turkmenistan, and the list goes on. They have all come together to discuss the resurrection of the economic corridor, known fondly as the Silk Road. Historically, this ancient network of trade routes connected East and West, and now the Chinese are seeking to breathe new life into it.

Little did I anticipate the surprise that awaits me. Suddenly, Hamid Karzai, the former president of Afghanistan, graces the stage with an aura of reverence and a symphony of thunderous applause. Ah, now I understand why the speakers' list was kept under wraps – security concerns, of course! Karzai delivers a poignant and eloquent speech, emphasising the values of peace, collaboration, and economic growth. Just as I rise from my seat, ready to join in the ovation, a man from Tajikistan accidentally whacks me on the head with his iPad while fervently snapping a photograph of Karzai. My poor headphones go flying, and my head throbs in protest. Yet, the man doesn't even muster a simple 'sorry'. Oh, the audacity! The rest of the morning drags on at a snail's pace, as one dignitary after another, representing the various '-stan' countries, along with Chinese Communist Party officials, regale us with speeches extolling the virtues of trade cooperation.

I anxiously glance at the time, realising that I must make a hasty escape to attend another meeting with journalists in the afternoon. I approach the first steward, seeking assistance, only to be met with unhelpfulness and an unwavering insistence that I board the bus with the rest of the delegates. Ordering a taxi seems like an impossible request in their eyes, and my frantic pleas fall on deaf ears. Seeking refuge from this transportation ordeal, I approach a second steward, only to be greeted with the same uncooperative demeanour, coupled with a generous dose of aggression. "Get back on the bus!" they bark, as if I'm an unruly child. The situation deteriorates further as I'm dragged onto the bus, scolded for my apparent defiance. The bus sets off, taking me even farther away from my intended destination. My heart races, and in a state of panic, I frantically call my colleagues, pleading for their rescue. Finally, the bus comes to a halt, and I'm granted permission to disembark, hailed as if it were a victory for the ages. It turns out others had faced the same transportation nightmare.

Alas, I arrive at the interview fashionably late and with a knock-on effect on the subsequent meeting at the university. Trapped in an endless sea of bumper-to-bumper traffic, my head throbbing from the earlier iPad encounter and my stomach grumbling from a lack of sustenance, I'm met with the disapproving gazes of the tutors. They are far from amused by my tardiness, ushering me straight into an auditorium filled with hundreds of eager accountancy students. As has become the norm at these events, a Communist Party official sits in the front row, scrutinising every word that escapes my lips and keenly monitoring the political correctness of the students' questions. Two seemingly never-ending hours pass by, and I find myself back in the clutches of the traffic log jam, my head throbbing, yearning for a moment's respite. Oh, the joys of an eventful day in Xi'an!

14 September 2013

As the sun rises, the wretched alarm blares once again, invading every nook and cranny of the hotel. It's as if we've been thrust into our very own version of *The Hunger Games*, minus the epic battles and heroic archery skills, of course. Fully awake and resigned to my fate, I opt for a chocolate-bar breakfast, embracing the unconventional start to the day, and diligently practise my presentation.

Imagine my utter disbelief when I arrive at the conference centre, only to be greeted by a desolate sight – empty conference rooms devoid of any delegates. It's as if the whole affair has been orchestrated purely for our benefit – the international speakers. Perplexed and disheartened, we exchange glances of disappointment, our travel efforts rendered futile. While we make do with presenting to one another, it feels like a hollow exercise, void of the intended purpose to share financial practices and developments with eager Chinese participants. It's a colossal waste of time, leaving us feeling like rather disgruntled bunnies.

The conference has turned out to be a clever misdirection, straight from the pages of the Chinese propaganda playbook. They cunningly seek to portray the city as a global leader, using our presence as a façade to demonstrate its allure to international businesspeople. We've unwittingly become pawns in their game, taken for a ride. This situation bears a striking resemblance to the previous conference I attended in Beijing, where I strongly suspect that most delegates were generously compensated to attend the first day of a three-day affair, purely to maintain appearances for us international visitors.

Oh, the trials and tribulations of navigating the unpredictable landscape of conferences in China. As the day unfolds, I can't help but wonder if there's an underground network of conference travellers swapping cautionary tales and war stories – tales of empty rooms, faux-delegates, and the eternal struggle to discern fact from fiction in this bewildering dance of international relations.

15 September 2013

As the dreaded screeching alarm pierces the air for the third consecutive day, I can't help but unleash a stream of expletives under my breath. Today, however, I have a different plan in mind – a budget-friendly adventure, coupled with the consumption of my last precious chocolate bar. Unlike my colleagues, who willingly succumbed to the trap of paying a whopping $200 to see the Terracotta Army,

I've cleverly devised a plan to venture there on my own, for a mere fraction of the cost. The challenge of navigating forty-two kilometres from Xi'an to the museum using local buses and taxis fills me with a sense of exhilaration.

Equipped with a bottle of water, I attempt to hail a taxi with all the finesse of a lost tourist, but to no avail. It's a less-than-ideal start to my grand adventure. However, fate takes pity on me as a passing motorcyclist kindly offers me a pillion ride, which I promptly decline on grounds of safety. Just when I begin to lose hope, the hotel concierge comes to my rescue, revealing the secret to successfully flagging down a taxi – boldly walk into the middle of the road and hurl oneself toward a passing cab. Who knew road-crossing acrobatics were a necessary skill?

Arriving at the bustling train station, I'm met with a chaotic scene – crowds of people queuing for tickets, scattered individuals occupying every inch of the concourse. As luck would have it, I have chosen one of the worst days of the year to travel as people are making their way home for the Mid-Autumn Festival, a national holiday when the entire country is in motion. The Mid-Autumn Festival, also known as the Moon Festival, holds immense significance for the Chinese, symbolising peace, prosperity, and family reunions. Every year, as the festival approaches, people embark on journeys across the country to join their loved ones in celebration.

Braving the sea of humans, I navigate my way to the bus stand, managing to secure the last available seat on the bus. The journey is nothing short of a bone-shaking experience, as we contend with relentless traffic jams, barely inching forward at times, until we finally break free from the clutches of the city. Passing through lush green countryside and mist-shrouded hills, the picturesque scenery is occasionally marred by the sight of countless single-use plastic bags dangling from trees. Upon closer inspection, I discover that they contain plastic-wrapped pomegranates, an unnatural and environmentally unfriendly sight. A fellow passenger enlightens me, explaining that the bags are used to artificially increase the size and weight of the fruit. Oh, the unsustainable lengths we go to in pursuit of plump produce!

We traverse small villages and towns adorned with vibrant red signage and flags until we reach our destination. As I disembark, my eyes are immediately drawn to the stalls selling fur and pelts of cats and dogs. It's a distressing sight for an animal lover, as a fur pelt resembling that of a German shepherd or a white Persian cat lies on display. In previous visits to China, I have been offered dog meat, which I vehemently declined. Unfortunately, the brutal dog meat trade still persists, and I can only hope that the government takes steps to end this cruelty in the future.

Entering the museum, I am awestruck by the sight before me – hundreds upon hundreds of terracotta warriors, horses, and chariots, meticulously arranged in battle formations dating back to 246 BC. Some warriors stand headless, while others are missing limbs, bearing the marks of ancient battles. Restoration efforts continue, as new pieces are continually unearthed, ensuring the preservation of this extraordinary historical site, which has rightfully earned its UNESCO status.

Under the scorching sun, I meander through the maze of shops amongst the haunting presence of deceased pets.

Upon this disconcerting sight, I stumble upon a shop selling intriguing antiquities. A figurine of a Maoist female farmer catches my eye, and the shop owner quickly seizes upon my interest, luring me into the depths of the shop. On bended knees, we peer through a curtain, revealing a hidden trove of antiquities. I can't help but feel that this illicit transaction would land the owner in trouble if the authorities found out. Offering yuan for the figurine, I soon realise that the shopkeeper's true interest lies in dollars, Swiss francs, Euros, or sterling. Thankfully, my purse contains a variety of currencies, and a deal is struck. With my newfound swag wrapped in an old newspaper, I embark on the rickety bus, beginning my journey back to Xi'an.

Arriving at the train station, I approach the taxi rank to negotiate a fare, only to be quoted an outrageous ¥500 for the short journey back to the hotel. This audacious sum is simply absurd, considering it had cost me a mere ¥100 to arrive. With polite resolve, I decline the offer and seek an alternative driver, who proceeds to repeat the same exorbitant quote. Mocking their outlandish demands, I inform them that they must be joking, knowing full well that my foreign protest will likely fall on deaf ears. Just as I'm about to walk away, the first driver approaches, flashes a smile, and reluctantly lowers his price to ¥300. I refuse to be taken for a fool and retort, "I'm no mug!" His smile fades, and in a moment of sheer elegance, he emits a guttural sound and expertly spits flehm onto the pavement – a common practice apparently considered healthy. A group of disgruntled taxi drivers now huddle together, shooting me menacing glances, resolute in their refusal to negotiate any further. At this point, I'm left with no choice but to walk away, my navigational skills haphazard at best.

Heat and dust envelop me as I trudge on, my weary feet protesting against each step. I find myself joining a dual carriageway, completely lost in the city. Desperation sets in as I spot a traffic policeman, desperately seeking directions. Yet, he dismisses my plea with a wave of his hand, leaving me stranded. Dehydrated and defeated, I reluctantly wave down a taxi, resigned to paying whatever exorbitant price is demanded. My will to live wanes, and true to form, I'm fleeced of ¥300 – the original sum I was quoted!

16 September 2013

Surprisingly, there is no blaring alarm to jolt me awake today. With my flight scheduled for the evening, I find myself blessed with some coveted free time to delve deeper into the wonders of the city. Stepping out into the scorching thirty-degree heat, I'm greeted by a haze of pollution that obscures the sun – a common sight in cities across Asia. Undeterred, I embark on a leisurely stroll towards the Xi'an Drum Tower, a magnificent relic from the Ming Dynasty, dating back to 1380. This historic structure served as both a timekeeper and an emergency alarm, signalling the passage of time or impending danger. Just as I arrive, the drummers take centre stage, unleashing a thunderous symphony of beats upon their cowhide drums. It's an unexpected fusion of ancient rhythms and dance music – a vibrant start to the day!

Next on my itinerary is the Bell Tower, located directly opposite the Drum Tower. Scaling the wooden

ARGENTINA

BANGLADESH

CHINA

HONG KONG

KOSOVO

MALDIVES

NEPAL

SOUTH KOREA

SOUTH AFRICA

tower, I find myself at the geographical heart of this ancient capital. Initially, the tower served as a lookout post, providing a vantage point to scan the surrounding countryside for any signs of aggressors. From this elevated position, I set off toward the Muslim Market, the bustling epicentre of the city's Muslim community.

The narrow streets of the Muslim Market are adorned with trees, casting dense shadows upon the scene. Within this labyrinth of tightly packed buildings, I find a myriad of restaurants, food stalls, and kitschy souvenir shops. Known as Beiyuanmen Moslem Street, or Muslim Street, this area was once home to foreign diplomats and merchants. Today, it is predominantly occupied by the descendants of those enterprising migrants, carrying on the tradition of selling their wares. As I venture further into the maze-like streets, I catch sight of men in white hats leisurely engaged in conversation on cane stools, while women shyly linger in the background, evading the camera's lens.

The market stalls are teeming with stacks of food, many of which I struggle to identify. Around every corner, tantalising aromas waft through the air, from spiced tea to freshly baked bread. The range of food on offer is astonishing – succulent beef, tender mutton and lamb, delectable buns, and mouth-watering fruit pies. Amongst the offerings, I spot a dish touted as one of their specialties – crumbled unleavened bread in mutton stew. Enticed by the fragrant scents, I'm even treated to a small cup of tea by one of the stallholders, providing a moment of respite to reflect upon the day's intriguing experiences.

In the final leg of my adventure, I ascend the towering walls that date back to the Ming dynasty, encompassing one of the most complete and expansive defensive systems in the world. The walls stand an impressive twelve metres tall and stretch fifteen metres wide, covering nearly nine miles of the cityscape. Finding a tranquil spot on the ramparts, I bask in the warm embrace of the sun's rays, momentarily distanced from the chaotic hustle and bustle below. A group of cyclists whizz by, reminding me of the missed opportunity to explore the entire circuit on a bike. Time ticks away, urging me to bid farewell to this enchanting city and make my way to the airport. Although the Euro Asia conference may have been a disappointment, these last two days have more than compensated for any initial let-down.

19. DRESSED TO IMPRESS FOR THE BENGAL CLUB, KOLKATA, INDIA

London Heathrow Airport –
Hamed International Airport, Doha, Qatar –
Netaji Subhash Chandra Bose International Airport, India

Flight Time: Twelve hours and thirty minutes

20 November 2013

As evening descends upon downtown Kolkata, I find myself clad in a tiger-patterned dress, oblivious to the hilarity it would elicit on this occasion. Little did I anticipate that my sartorial choice would become a source of amusement as we arrive at the venerable Bengal Club – the oldest social and business club in India, with a rich history dating back to 1827. Throughout the years, luminaries from Mahatma Gandhi to Rudyard Kipling have graced its esteemed halls.

Upon stepping into the minibus, swathes of laughter erupt from my male colleagues, who seem convinced that I intentionally dressed to resemble a Bengal tiger. Oh, the unintended consequences of my fashion choices! In that moment, I couldn't help but envision myself, dress and all, mounted as a trophy above the club's entrance hall.

The jokes persisted throughout the evening, fuelled by copious amounts of Black Dog Scotch whisky imbibed by my mischievous male counterparts.

21 November 2013

Today is a whirlwind of meetings intertwined with a visit to the serene grounds of Calcutta University, a bastion of academia amidst the bustling chaos of the city.

Kolkata, the capital city of West Bengal (formerly known as Calcutta), exudes a duality that captures my attention. Rickshaws share the roads with luxurious cars, beggars perch on the walkways while well-dressed individuals stride past. The affluent relish in air-conditioned comfort and dine at five-star restaurants, while the less fortunate make do with food prepared in humble wooden shacks.

With some spare time after exploring the university, I hail one of the iconic yellow taxis that epitomise Kolkata's character. We navigate through the chaotic traffic, eventually arriving at the majestic Victoria Memorial – an imposing white marble edifice paying homage to Queen Victoria. Surrounded by peaceful gardens, it stands as a museum where I indulge in a leisurely stroll.

Returning to the taxi, we zigzag through the congested streets, ultimately reaching our next destination – Mother House. This sacred abode shelters a Roman Catholic congregation established by Mother Teresa in 1950, dedicated to caring for the poor, abandoned, and homeless. Arriving early, I find myself standing alone in a labyrinth of narrow alleyways, surrounded by some of Kolkata's neediest individuals. Beggars approach me, seeking solace in monetary relief.

The sight is heart-wrenching. Adults and families sleep side by side on the walkways, while young children, as young as two, wash themselves in the dirty water flowing from the gutters onto the road. They face the constant danger of speeding traffic, coupled with the risk of waterborne diseases. Mother House remains as relevant today as it was in 1950 – poverty continues to afflict Kolkata, with stark disparities in living conditions, healthcare, and education within the slums.

Feeling somewhat vulnerable outside Mother House, awaiting entry, I am relieved when a passing nun guides me inside. She leads me to a small museum that sheds light on the religious order and its missionaries. Another section is dedicated to Mother Teresa herself, housing her tomb alongside an exhibition showcasing her work and personal belongings. It's impossible not to be emotionally moved by the selflessness and compassion she embodied.

The day has been long, yet the evening holds an unexpected delight. Unbeknownst to me, we embark on a cruise along the Hooghly River, a vast expanse of water that cuts through the city. Alas, my choice of attire once again proves impractical – I am donned in a dress that exposes my arms and legs, much to the delight of the voracious mosquitoes swarming the area. Their itchy bites and omnipresent buzzing threaten to overshadow the evening, and I find myself engaging in a futile battle to keep them at bay.

As I gaze over the side of the boat, I witness the remnants of funeral pyres and delicate flower petals drifting down the river. The distant riverbank reveals the fiery glow of cremations – a solemn practice in Hinduism, as open-air burning is considered spiritually appropriate for releasing the soul from the body after death.

As the boat navigates under the colossal Howrah Bridge – a behemoth of steel suspended over the river – the unexpected unfolds. The air fills with acid house music, igniting an explosion of revelry on deck. Even our usually composed host succumbs to the contagious energy, hanging from the mast with his shirt flung open, gyrating with abandon. Spontaneously, others join in, unleashing a frenzy of dad dancing and twirling like ecstatic dervishes. Just when we believe the boat is nearing its destination, it veers around, caught in a loop. The music intensifies, the dancing grows wilder, and the Black Dog Whisky makes its triumphant return. The revellers are drenched in sweat, providing an irresistible feast for the relentless mosquitoes. We make three circles without any sign of disembarking, until finally, exhausted, we are granted permission to step ashore.

20. THRILLER IN MANILA

London Heathrow Airport –
Ninoy Aquino International Airport, Manila, Philippines

Flight time: Thirteen hours and twenty five minutes

25 March 2014

You might call me crazy, but here I am, embarking on a journey halfway across the globe, spanning an impressive 10,734 kilometres, all for just two measly working days. It's one of those trips that defies the norm, where immersing myself in the local culture is a mere impossibility. With my bags in tow, I board the plane, preparing myself for a gruelling fourteen-hour flight.

Upon landing, the wearying saga continues as I find myself trapped in standstill traffic. The driver, with a sigh, informs me, "It might be hours before we reach the hotel, despite it being just a stone's throw away." Exhaustion takes hold of me, and all I can think of is finding solace in the sanctuary of my hotel room, attempting to combat the formidable enemy – jet lag.

These working days offer little to jot down in the annals

of memorable experiences. I find myself navigating through predominantly male-dominated meetings held within freezing air-conditioned rooms, followed by a monotonous seminar. Jet-lagged participants, hailing from different corners of the world, struggle to keep their eyes open, their energy depleted. The conference suite transforms into a scene straight out of a somnolent symphony, with glazed eyes, heads bowed on desks, and some even finding solace in an open slumber. The lack of questions and interaction with presenters becomes a stark reality, with fatigue reigning supreme. A few individuals resort to sleeping pills in a valiant effort to counteract the merciless effects of jet lag, but I've never been one to venture down that path.

However, amidst the drudgery, there are small consolations – our host, the Asian Development Bank, graces us with the splendour of a beautiful and ornate marble building, reminiscent of its noble mission to provide aid to developing countries. And as evening descends, I stumble upon the hotel's rooftop cocktail bar, a haven that offers a modest respite from the mundane.

Sometimes, in the whirlwind of business travels, the memories that truly linger are not of the conference rooms or the hectic schedules, but rather the fleeting moments of reprieve, the sights and sounds that momentarily transport us beyond the realm of spreadsheets and PowerPoint presentations.

25 March 2014

As the day unfolds, the clock mercilessly ticks away, leaving me with precious little time to venture beyond the confines of the hotel before bidding this city farewell. Today's adventure, or lack thereof, will be etched in my memory as I find myself seated amongst my colleagues in a humble takeaway joint nestled within the cavernous depths of a gargantuan shopping mall. To say that none of us were thrilled with the restaurant choice would be an understatement. It became abundantly clear that the culinary offerings consisted mostly of fried fare, leaving us yearning for something more satisfying. To add insult to injury, our quest for a simple bottle of wine turned into a mission as we patiently waited for the owners to embark on their impromptu shopping excursion.

Despite the lacklustre ambience and our collective state of fatigue, we tried our best to make the most of the situation. Tense and weary, we shared forced smiles and engaged in polite conversation, attempting to salvage whatever enjoyment we could muster from this unconventional dining experience.

Oh, what a whirlwind this short trip has been! As I embark on the journey back home, a sense of relief washes over me. The memories, however fleeting and unconventional, will forever be etched in my mind. And now, with a mix of exhaustion and anticipation, I step onto the homeward-bound flight, bidding farewell to the city that has offered both challenges and fleeting moments of respite.

21 BARBADOS, HERE I COME!

**London Heathrow Airport –
Grantley Adams International Airport, Barbados**

Flight time: Nine hours and forty-five minutes

23 September 2014

As I bask in the golden sunlight the infectious melody of the song 'Barbados' by Typically Tropical from 1975 sticks in my mind. It's the perfect soundtrack to accompany my leisurely respite on these picturesque sands, a well-deserved break after two intense days of chairing financial management workshops with government officials.

Amidst the serene beauty, a red warning flag flutters in the breeze, signalling treacherous currents that pose a danger, particularly for children. However, there always seem to be those who either fail to notice or deliberately ignore the warning signs. Just as I observe, a young child ventures into the water, only to be abruptly knocked over by a powerful wave. His neglectful parents rush to his aid, but not before he sprains his ankle. Sometimes, the folly of human behaviour knows no bounds.

Nevertheless, Barbados lives up to its well-deserved

reputation as a paradise, with its pristine beaches, windswept bays, and captivating flora. This afternoon, I find myself visiting the enchanting Andromeda botanic gardens, nestled along the west coast. Every turn reveals a delightful surprise, making it the ideal place to unwind and soak in the natural beauty. These gardens, a testament to the creativity of a remarkable Barbadian woman named Iris Bannochie, who was a leading horticulturalist, serve as a true testament to the wonders of nature.

Like many other Caribbean nations, Barbados boasts a representative democratic government. A key purpose of my trip was to meet with public officials from the Ministry of Finance in Bridgetown, the vibrant capital, and lend support to their efforts in improving financial accounting practices. As a perpetual student of democratic governance, I eagerly accepted their invitation to tour the parliament buildings, which house both the assembly and the senate. These two blocks of architectural grandeur encapsulate over 375 years of democratic heritage and history. Portraits of British sovereigns from James I to Queen Victoria adorn the walls, alongside past presidents of council and speakers of the assembly. However, it was the intricate stained-glass windows along the main stairway, adorned with biblical quotations, that truly captivated me. They served as a poignant reminder of the profound influences that shape a nation's governance.

With the tour of the parliament buildings serving as the perfect finale to a fulfilling day, I can't help but revel in the rich cultural tapestry and historical significance of this remarkable island. Barbados truly offers an idyllic blend of natural beauty and democratic spirit, making it a destination that leaves an indelible mark on the heart and mind.

24 September 2014

Bright and early, I find myself at the airport, ready to embark on another journey, this time to Kingston, Jamaica, for a series of workshops. As I check in, it dawns on me that many people underestimate the vast distances between Caribbean islands. Unlike the notion of conveniently hopping from one island to another, the reality is far different. The journey from Barbados to Jamaica spans a staggering 1976km, with a flight duration of three hours. So, despite the relatively short distance on a map, it will consume the better part of my day to reach my destination, door to door.

This isn't my first encounter with the challenges of travelling between Caribbean islands. On a previous trip, I ventured from Jamaica to Trinidad and Tobago, covering an equally considerable distance. Such journeys require meticulous coordination and careful planning to ensure that multiple business trips within the Caribbean region are both efficient and worthwhile.

As I settle into my seat on the plane, I reflect on the effort it takes to navigate these vast distances and appreciate the unique logistics that come with conducting business in this part of the world. The Caribbean, with its diverse array of islands, presents both opportunities and challenges, but the chance to engage with different cultures and contribute to the region's development makes it all worthwhile.

22. JAMAICA – COFFEE BEANS AND INDEPENDENCE

Grantley Adams International Airport, Barbados – Norman Manley International Airport, Jamaica

Flight time: Three hours

2 September 2014

After a couple of days filled with never-ending meetings and workshops, I finally have a free day before my journey back home. Today, I've arranged to visit a coffee plantation in the stunning Blue Mountains with the help of George, the driver I met during my previous visit. Accompanying me is Harry, an avid Arsenal football fan who's bursting with enthusiasm and lively conversation. We spend the entire journey chatting and giggling about life, crime, world politics, and, of course, the Premier League football.

As we traverse the narrow, steep, and winding roads, we pass charming villages and roadside shacks peddling tourist trinkets. The scenery becomes increasingly mesmerising as we arrive at our destination, the Crighton Coffee and Tasting Farm, nestled high in the Blue Mountains.

Due to our timing, we've just missed the first tour and are asked to wait until Joseph, the guide, returns. Although the wait is a tad frustrating, it offers me the chance to sample some smooth Arabica coffee, known for its sweetness and purported heart-health benefits. Just as we're about to embark on the tour, Joseph reappears, accompanied by an American tourist from Seattle who shockingly declares that he hates coffee and hasn't tasted it in over thirty years. Understandably, he won't be purchasing any to take home. It's a peculiar sight to see him visiting a coffee plantation despite his clear aversion to coffee, but I can relate, having visited numerous whiskey distilleries despite not enjoying the taste of whiskey. It takes a while to bid him farewell, which irks me as it delays my own tour.

Finally, Harry, Joseph, and I begin our ascent up the rocky paths of the coffee plantation. Along the way, Joseph shares insights about the history of coffee cultivation in Jamaica, the inner workings of a coffee farm, and how their coffee has garnered a reputation as one of the world's most expensive. The vistas of lush green hills in the distance are awe-inspiring, and I find myself absorbing a wealth of knowledge about the world of coffee. After a breathless climb, we reach the summit and eagerly take a much-needed rest.

It's here, surrounded by the serene beauty of coffee bushes and basking in the warm sun, that our conversation takes an unexpected turn. We discuss a range of topics that extend far beyond coffee cultivation. I'm taken aback when the subject of the Scottish Independence referendum arises. While it's a current and significant topic in the UK, I'm surprised to find two young men in Jamaica expressing an interest. To my astonishment, both Harry and Joseph firmly believe that Scotland should remain part of the UK and go so far as to jokingly accuse 'leave' voters of having mad cow disease. In an impromptu poll amongst the three of us, hands are raised unanimously in favour of Scotland staying within the UK. Intrigued by their perspective, I pose the question: "What are your thoughts on Jamaica's fifty-three years of political independence from the UK?" Once again, I'm taken aback as Joseph replies, suggesting that the country made a premature move toward independence and should have waited until its political structures were more mature.

Surrounded by the vibrant greenery of coffee bushes and filled with laughter, our animated discussion attracts the attention of a group of tourists passing by, who decide to join in on the fun.

As we make our way back down the mountains, Harry and I continue our banter, laughing heartily all the way back to Kingston.

23. OVERDRESSED, WASHINGTON, DC

London Heathrow Airport –
Washington Dulles International Airport, US

Flight time: Eight hours and fifteen minutes

14 October 2014

After delivering a talk on integrated reporting, an important reporting requirement for governments to showcase their policy impacts on the environment, economy, and society, Helen from the World Bank invites me to join her for lunch at the bank's café. I had anticipated a grand dining room to match the World Bank's status, but to my surprise, it resembled an ordinary office canteen, lacking any distinctiveness.

Over a delightful lunch, Helen and I engage in a conversation about politics, and somehow the topic veers towards the latest technologies for tracking cats and dogs. Helen is amazed when I mention that the welfare of dogs is taken very seriously in Buenos Aires, evident from the abundance of high-end pet pampering shops. She laughs and adds, "That's nothing compared to the US, where pet indulgence reaches astronomical levels. Dogs are not only assigned dog walkers who cost $500 a month, but they also sport GPS trackers, and their owners even pack them lunches before heading off to work." She shares a story about a dog walker who was fired when the dog's GPS tracker revealed that it hadn't left the house all day, despite the walker's insistence of taking it on a long walk.

Putting our lunch conversation aside, I feel immensely privileged to have had the opportunity to speak at the World Bank, an organisation composed of members from numerous countries striving to reduce poverty and foster shared prosperity through technical and financial assistance. Many colleagues flew in to attend the event, and to my surprise, Joe, a former colleague I hadn't seen in years, showed up at the bistro in downtown Washington, DC last night. It was a pleasant surprise and little did I know that I would also meet Mitzi, a leadership development consultant, for the first time, and she would become a dear friend for years to come.

Back at the hotel, I engage in conversation with Gabrielle, a US civil servant attending a government training course. Over a glass or two of wine, we exchange experiences about the US and UK governments. Gabrielle mentions that it's an intriguing time to be in Washington as there is speculation about the possibility of a government shutdown due to the failure to agree on a budget. I enquire whether this is a regular occurrence, to which she replies, "It was just over a year ago that the government shut down for sixteen days when Congress couldn't reach an agreement on a budget for the new fiscal year. Nearly 800,000 federal employees were furloughed without pay, and an additional million experienced pay cheque delays."

She explains that it all started when Obama submitted his budget to Congress, which failed to agree on a budget for the Affordable Care Act, also known as Obamacare, aimed at reducing healthcare costs for families. She emphasises that this year the disagreements revolve around immigration. Unlike the UK's Westminster system, where budgets are typically approved, the US Congress has the option to reject a budget.

15 October 2014

Visiting Washington, DC, the renowned city that serves as the seat of the US Federal Government and hosts numerous global charities and international development institutions, is a powerful experience. As I browse through the World Bank bookshop, I stumble upon *Overdressed* by Elizabeth Cline, a captivating book that sheds light on the environmental cost of cheap fashion.[9] It's hard to put the book down as it reveals the staggering amount of garments we consume and hoard each year, contributing to an environmental crisis while the planet's resources deplete. The revelations in the book make me regret not having read it earlier, as it could have provided valuable insights for my presentation.

Eventually, I tear myself away from the book and the mounting work emails to embark on a city walk. Passing by government buildings, I'm struck by the immense political power and influence concentrated in this city. Along my route, I encounter significant landmarks such

---

9   Elizabeth L Cline, *Overdressed: The Shockingly High Cost of Cheap Fashion*, 2013

as the Lincoln Memorial, a tribute to Abraham Lincoln, alongside a reflecting pool mirroring the vibrant autumn foliage. I also pass by the Vietnam Veterans Memorial, a solemn site engraved with the names of soldiers who lost their lives during the Vietnam War, and the Marine Corps Memorial honouring the bravery of marines in various conflicts, including World War II.

Despite the fatigue and soreness in my feet from walking on concrete pavements, I press on towards the White House. Along the way, I encounter several homeless individuals wrapped in layers of clothing, seeking warmth from the heating vents on the sidewalks. It's disheartening to witness such poverty outside the symbol of power, especially in a country where billionaires' combined net worth surpasses the GDP of many states.[10]

Arriving at the White House, I'm surprised by its relatively small size compared to grand mansions on London's 'Billionaires' Road'. Knowing that President Obama isn't present, I accept that I won't catch a glimpse of him on the famous lawn. However, just around the corner, I stumble upon a photo booth designed to resemble the Oval Office. Unable to resist, I take a photo that gives the illusion of delivering a speech behind a lectern in the Oval Office.

Returning to the hotel, I delve back into the book on fast fashion, and it prompts me to make life-changing decisions. I resolve to reduce my clothing expenditure, resist the allure of discounts, and prioritise recycling. Letting go of the temptation to buy cheap clothes from

---

10 *https://247wallst.com/state/this-is-the-city-in-washington-with-the-most-billionaires/*

China, which accounts for 90% of our clothing production, will undoubtedly pose a challenge. Nonetheless, I am determined to make a positive change and contribute to a more sustainable future.

24. RUNNING UP THE SPANISH STEPS

**London Heathrow Airport –
Leonardo da Vinci International *Airport, Rome, Italy***

Flight time: Two hours and twenty minutes

9 November 2014

It's exciting to be in Rome, where accountants from around the world are gathering for the prestigious World Congress of Accountants. With nearly five thousand attendees expected, the event is often referred to as the 'Olympics' of accountants, providing a platform for thought leadership and global practice exchange in the field of accounting. As I arrive to support the team organising a fringe event and an early evening reception, the hotel is abuzz with accountants, and their presence is inescapable.

Although I have explored Rome on previous visits, the busy schedule of the congress means there is less time to indulge in sightseeing this time. While I will miss out on delving into the city's ancient history, visiting world-class museums, and experiencing iconic landmarks like the Roman Forum and the Colosseum, I do plan to seize moments to savour a few espressos in the beautiful piazzas.

Despite the limited opportunities for exploration, being part of such a significant accounting event in the eternal city is an exhilarating experience in itself.

10 November 2014

Just a couple of months ago, I accomplished a personal goal: completing the renowned Great North Run, a popular half marathon held in the vibrant city of Newcastle Upon Tyne. It stands as a significant achievement, and I take great pride in crossing that finish line. Since then, running has become a cherished part of my routine, a joyful activity without the pressure of meeting specific distance targets.

As the first rays of sunlight break through the gentle morning mist, I lace up my trusty running shoes and step out onto the serene streets of Rome. My intention is not to push myself to the limit but rather to indulge in a leisurely jog, soaking up the beauty of the city. I begin by descending down winding streets, savouring the charm of each narrow path, until I find myself passing by inviting piazzas and elegant rows of luxurious shops. Eventually, I arrive at Piazza di Spagna, standing at the foot of the historic Spanish Steps – a renowned landmark that dates back to the sixteenth century.

135 steps lie before me, leading to the summit, and I find myself hesitating. Will my fitness carry me all the way to the top? Uncertainty lingers, but a spark of determination ignites within me. I decide to embrace the challenge and embark on a steady and focused ascent. With each step, I grow closer to my goal. Before long, I reach the pinnacle, catching my breath as I gaze out at the magnificent views

that unfold before me. The sun rises higher, casting its warm rays upon the city, and a sense of achievement washes over me. This simple pleasure of conquering the Spanish Steps and witnessing the beauty of Rome is my gift to start the day.

In this moment, I'm reminded of the iconic scene from the 1976 film *Rocky*, where Sylvester Stallone runs up the stone steps leading to the entrance of the Philadelphia Museum of Art. Just like Rocky, I've conquered my own personal challenge, ready to face the 'Olympics' of accountants with determination and fortitude.

25 NORTHERN IRELAND – CAREFUL WHAT YOU WEAR

London Gatwick Airport – Belfast International Airport, Northern Ireland

Flight time: One hour and twenty five minutes

14 February 2015

It's a frantic morning as I rush out of my front door at 6am, aiming to catch a train to Gatwick for a flight to Belfast. I have an important meeting in Stormont, the Northern Ireland Assembly. But as I reach the bus stop, panic sets in. I realise that I'm wearing an orange dress. In the Irish context, the colour orange carries political and symbolic connotations, often associated with the Ulster Unionist (UUP) and Democratic Unionist (DUP) parties. I can't risk offending the Irish Nationalists by walking around Stormont in orange. Unfortunately, I don't have time to go home and change, as it would cause me to miss my flight. Desperately, I hope that a women's clothing shop at Paddington station will be open to save me from this fashion disaster. However, luck is not on my side and the shop is closed. I have no choice but to improvise by covering up the orange dress with a buttoned jacket and a scarf.

During the short flight, I resort to googling 'orange and Irish politics' to ease my worries. Luckily, I discover that wearing orange-coloured clothing holds no significance unless it's accompanied by a sash and a bowler hat. Thankfully, I have no intentions of donning such an outfit.

Upon arrival, I'm greeted by John, a wise civil servant, who accompanies me to Stormont. As we drive along the tree-lined avenue leading to the parliament, John informs me that there are over three hundred lime trees adorning the route. Stormont itself is an impressive building, housing the Northern Ireland Assembly – the legislative body. It's hard to fathom why it was once described as a 'large plain house'.

I keep my scarf draped discreetly to conceal the orange dress as we navigate the lengthy corridors and make our way to meet the Minister of Finance, who happens to be from the DUP. As a democratic unionist, I hope my choice of colour will be well received by him.

I find myself in Northern Ireland during a time when the government operates on a power-sharing agreement. This arrangement ensures representation for both nationalist and unionist communities in Parliament. The concept behind power sharing is that despite historical differences, both communities have a vested interest in making the system work. However, this system has been fraught with difficulties over the years, leaving Northern Ireland without a functioning government. A power-sharing arrangement returned in 2024.

To my surprise, I encounter a Minister of Finance who demonstrates a genuine interest in finance and a willingness to learn from other countries worldwide.

It's a refreshing change to see a government looking beyond its borders. While I don't expect politicians to be accountants, it's not unreasonable to anticipate a certain level of financial acumen in parliaments, especially when budget decisions are at stake. I recall a rather disheartening incident in the House of Commons where an MP asked his researcher to calculate a simple VAT percentage on a purchase. Engaging in an informative exchange about international budgeting practices, I prepare for a tour of the parliamentary estate.

As we navigate the grandiose corridors, fascinating facts about the architect, Arnold Thornely, come to light. His attention to detail is remarkable, ensuring that the length of the building measures exactly 365 feet, symbolising one foot for each day of the year. Inside the central part of the building, a magnificent hall greets us, adorned with six pillars crafted from Italian marble and vibrant, painted ceilings. Countless significant events have taken place here, including the funeral of George Best, a renowned footballer. As we sneak into the public gallery of the assembly chamber, we even catch sight of a former IRA member, implicated in a series of bombings in the 1970s and 80s, who has now become a Sinn Féin politician.

It has been a long and eventful day, and I successfully managed to conceal the orange dress. As we relax in the members' room, John poses a thought-provoking question. "Gillian, do you fully comprehend the significance of the colour in Northern Ireland politics?"

I respond, "I do now." It's a valuable lesson learnt – pay closer attention to dress attire in the future.

26. HEADING FOR THE UNITED NATIONS, GENEVA, SWITZERLAND

**London Gatwick –
Geneva International Airport, Switzerland**
Flight Time: One hour and forty minutes

16 March 2015

This morning, I find myself boarding a flight to Geneva for a speaking engagement at the Palais des Nations, the United Nations headquarters. It's my first visit to Geneva, and I'm filled with anticipation to see the iconic art-deco building that once served as the home of the League of Nations in the 1930s. The League of Nations was a collective effort by countries worldwide to resolve disputes after World War I, albeit with mixed success. Following the devastating aftermath of World War II, representatives from fifty nations gathered and crafted the UN Charter, a pivotal document that gave birth to the United Nations. The hope was to prevent another global conflict and strive for lasting peace.

Upon my arrival, I find myself with some spare time, so I embark on a journey to explore the hidden corners of this renowned hub for conference diplomacy. As I glance to my right, I'm captivated by the sight of the beautiful art-deco building that houses the United Nations. Its grand entrance is adorned with vibrant flags representing nations from all over the world, gracefully fluttering in the breeze, beckoning towards Lake Geneva. I eagerly anticipate stepping inside the building tomorrow. To my left stands the headquarters of the Red Cross, an organisation that has played a significant role in humanitarian efforts.

As I stroll along the right bank of the lake, I'm treated to astounding views of the majestic Alps. In the midst of the lake, the magnificent Jet d'Eau stands tall, proudly boasting the title of the world's tallest water fountain. Taking a moment to recharge, I make a quick pit stop for a cup of coffee before delving into the enchanting old town. Its winding cobblestone streets and the historic Town Hall, which once housed the League of Nations and the Red Cross, create a charming and intriguing atmosphere. After a relaxing day of exploration, I retreat to the comfort of the Bijou Hotel to prepare for the conference that awaits me the following day.

17 March 2015

As expected, entering the UN premises involves thorough security checks. Passport details are exchanged and carefully recorded, followed by the obligatory photographs, before being escorted to one of the numerous conference rooms.

Moments before my presentation, a wave of nervousness washes over me. To combat it, I take a deep breath and mentally rehearse the contents of my slides.

I remind myself of the privilege it is to be standing here. Not many individuals have the opportunity to present in such a prestigious setting. Throughout my career, I have experienced various milestones, from working in the House of Commons to supporting supranational bodies. Yet, this moment feels like another pinnacle.

Despite the initial jitters, it appears that my presentation has resonated with the diverse audience hailing from Mexico to the Netherlands. I feel a sense of relief now that it's concluded, allowing me to finally sit back and attentively listen to the other speakers. However, there's a minor hiccup: there isn't a nameplate designated for 'Great Britain'. Without hesitation, I opt for the nameplate of 'Seychelles'. It strikes me as amusing and somewhat absurd, dubbing myself 'Miss Seychelles' for the day. The light-heartedness brings a smile to my face.

27. TOKYO REVISITED

**London Heathrow Airport –
Narita International Airport, Tokyo, Japan**

Flight time: Fourteen hours and ten minutes

27 May 2015

The time difference between Tokyo and London is a real killer when it comes to recovering from jet lag. My body clock is in overdrive, just like it was during my first trip to Tokyo a decade ago. That visit was quite the adventure, filled with visits to the Tokyo Metropolitan Council and local governments, along with a homestay experience. I learnt so much about Japanese culture and their way of working, although back then, there were few Japanese women in senior positions. They were often pressured to marry at a young age, and their careers would often come to a halt after having children. In the meetings I attended, I noticed that the few women present were usually seated at the back of the room, and as a young female professional, I sometimes felt like a bit of a puzzle to the government officers. I always seemed to be at the end of the line for questions.

One would hope that gender equality has made progress since then, but the evidence suggests otherwise. Japan still grapples with a significant gender pay gap, one of the highest amongst developed countries. Women's political participation and empowerment also lag behind their neighbouring countries.[11] It's disheartening to see the challenges that persist.

During this current visit, my purpose is to share good financial practices and developments through a series of seminars and events with finance colleagues from Asian governments. I'm hoping that the cultural etiquettes I learnt on my previous trip will come in handy during our planned meetings. Take, for example, the ritual of exchanging business cards. In Japan, this practice holds much greater value and significance than in Western countries. It's an elaborate ceremony based on respect for order and rank, with a strict hierarchical order dictating the card exchange. Those in higher-ranking positions take the lead, and it works its way down to the lowest position. The most senior person is usually the oldest, too.

As I sit in the meeting, my mind wanders to other cultural etiquettes I encountered during my homestay. I had to remember to wear slippers in the house, and when using the toilet, I had to change into specific toilet slippers and change back before walking around the rest of the house. Thank goodness I remembered and didn't commit any heinous crimes! I was also surprised when I was offered a set of clean towels and directed to the bathroom.

---

11   https://stats.oecd.org/index.aspx?queryid=54751

It wasn't because they considered me a dirty alien, but rather it was a social custom and a kind gesture.

Bathing had its own set of rules as well. I had to be careful not to let the wash towel or any soap enter the bathtub, and I wasn't the only one using the same bathwater. It was quite overwhelming to remember all these rules, but I did manage to become adept at using chopsticks and enjoyed sleeping on a traditional Japanese silk floor mattress set on tatami mats.

One cherished memory from a previous visit was spending time with Ichika, my host for a weekend. She organised a girl's evening at home with friends who wanted to practise their English. To my surprise, I was introduced as an 'alien', a term often used to describe international visitors. They eagerly asked me about every aspect of my life while sipping from tiny glasses of red wine, which didn't seem to be their preferred choice, but they drank it to make me feel welcome. The big news of the day for them was the discovery of a mutilated body of a Western woman in the Roppongi area, where she had gone missing. They were curious to know if this gruesome news had affected me since my arrival in the country and whether I felt safe. Ichika's friends left early, and the evening was a stark contrast to a wild and raucous girls' night out in the UK.

As I reminisce about these memories, our business meeting abruptly comes to an end, and a few of us head to the rooftop bar, overlooking the Tokyo skyline with its dazzling neon lights and towering skyscrapers. As I comfortably sit at the cocktail bar, I can't help but be reminded of the film *Lost in Translation*, the acclaimed

2003 comedy where a wide-eyed whisky-drinking Bill Murray meets Scarlett Johansson in a hotel bar and forms an unlikely bond. I half expect him to walk in at any moment!

28 May 2015

The meetings are proceeding at a rapid pace today when suddenly I feel a jolt followed by a stronger shaking. The whole building trembles, and the grand glass chandeliers in the lobby start swinging violently, prompting the staff to quickly guide everyone away, just in case they come crashing down. The earth tremor lasts for a brief but intense two minutes, leaving people scared and shocked. Some of my colleagues rush into the lobby looking pale, as they had been trapped in a lift on the upper floors when the tremor began. They managed to escape by prying open the doors and dashing down the emergency stairs.

Earthquakes are not an uncommon occurrence in Tokyo, and having experienced them before, I'm less worried than my colleagues. The Japanese meteorological agency reports around two thousand earthquakes striking the area each year. It's comforting to know that the buildings are constructed to withstand such tremors.

As I make my way to Narita Airport, I gaze out of the window at the vibrant, brightly coloured lights that illuminate this bustling metropolis of over 38 million people. Tokyo truly comes alive at night, and it reminds me of the times I rode the JR line circling the city to discover its only Cambodian restaurant or wandered through the labyrinthine streets of Shinjuku, hunting for great deals on electronic goods. I have a fondness for Tokyo, and I'm certain that I will return to this vibrant city at some point in the future.

28. RETURN TO KENYA – THE BODYGUARD AND ME

**London Heathrow Airport –
Jomo Kenyatta Airport, Nairobi, Kenya**

Flight Time: Eight hours and fifty-nine minutes

6 March 2016

I can't help but question my decision to book a flight that lands in Nairobi airport at 2am, especially with the UK foreign office warning of potential attacks in such places. Arriving at this ungodly hour feels like an unnecessary risk, even though I've been to Nairobi before.

To my surprise, the conference organisers are still here to welcome me. I'm introduced to Mwangi, a robust and imposing figure who will be my bodyguard for the next few days, my very own James Bond. We exchange handshakes before I'm swiftly escorted into a car, and we drive off into the darkness. Nairobi lacks streetlights and reflective road markers, making the journey quite unnerving as we navigate through pitch-black roads. I hope Mwangi is indeed a genuine bodyguard and not some sort of kidnapper!

I had made an online booking for a hotel, but I knew little about its location. As we enter Bishop Road, a secluded area of Nairobi, we're met with large concrete bollards and ribbons of barbed wire. Soldiers stand guard with machine guns slung over their shoulders. None of this was mentioned in the hotel description, nor was it pointed out that the hotel is situated next to the highly fortified Israeli Embassy. However, it turns out that this unexpected security detail makes it one of the safest places in town, and surprisingly, it's neither shabby nor exorbitantly priced like most hotels in Nairobi. Plus, I had read that it boasts a classy brasserie and wine bar, reputed to serve some of the best sushi in the city. So, at least there's something to look forward to.

I'm waiting for Mwangi to arrive and escort me to the venue. Just as I'm about to lose hope, a UN vehicle pulls up with Mwangi and his driver. He greets me with a wide smile, saying, "Good Morning, Ms. Fawcett," before we set off. As we drive through the city, I notice that it's similar to most cities on a weekday morning. The streets are bustling with commuters making their way to work, except here it seems most people are walking rather than driving. A swarm of briefcase-toting individuals in crisp white shirts move in unison. We arrive at the conference centre, and I'm efficiently guided inside. Despite being on time, I join other participants in impatiently waiting for the conference to commence. In my experience, you can never be certain that a conference in Africa will run exactly as planned.

Throughout the day, Mwangi's attention to detail is unparalleled. If I need a pen, he swiftly provides one. If I require a glass of water, he already has it prepared. And if, hypothetically speaking, I were to request a vodka martini, well, let's just say he would rise to the occasion. I'm only joking, of course. During each session I attend, Mwangi remains stationed outside the room, occasionally peering through windows and around corners to ensure my safety and keep a watchful eye for any suspicious activities. With around one thousand attendees at the conference, I'm never out of his sight, perhaps because I'm the sole attendee with blonde hair. And so ends a very long day as I'm transported back to one of the safest places in the city.

7 March 2016

I'm promptly dropped off by the driver at the conference this morning, where I deliver a presentation on financial management. Afterward, I have a few hours to relax and listen to presentations from other speakers. Following a satisfying lunch, Mwangi is relieved from his duties, and I'm entrusted to the capable hands of Jack, another driver who will accompany me to one of Nairobi's most popular attractions – the Sheldrick Wildlife Trust. This renowned trust is known for its successful efforts in rescuing and rehabilitating orphaned elephants.

As we drive towards the trust, Jack shares some fascinating information with me. He tells me that the trust was founded in 1977 by a remarkable woman named Dr. Dame Daphne Sheldrick, in memory of her late husband. Even today, the trust is operated by her family, carrying on her legacy. It's heart-warming to see the elephants, a majestic herd of bronze giants, emerge from the bushes, excitedly anticipating the lunchtime treats.

The trust serves as a beacon of hope for these vulnerable creatures, who face threats from poachers, habitat loss, deforestation, and drought. The work they do to ensure the survival of orphaned elephants is truly remarkable. However, reading the information boards, I'm reminded that not all of them make it through the critical recovery phase. Some have endured the loss of their mothers and herds, leaving them deeply traumatised with both physical and psychological scars. The battle to overcome the shock and grief of their lost loved ones can sometimes tip the scales of their survival.

The Sheldrick Wildlife Trust stands as a sanctuary, providing these incredible animals with a chance at a better future. It's a humbling and emotional experience to witness their journey and the tireless efforts of the dedicated individuals who care for them.

8 March 2016

A torrential downpour is brewing in Nairobi, threatening to bring the city to a complete standstill. My anxiety levels rise as I realise I need to make it back across the city from the conference centre to the hotel to collect my bags, and then navigate through the chaotic streets once again to reach the airport. Mwangi, my trusty bodyguard, receives a call from his driver who informs him about being stuck in a traffic jam just two miles away. It feels like everything is going wrong!

Spotting a stationary police car nearby, Mwangi springs into action and manages to secure us a ride. Now it's me, two police officers, and my bodyguard squeezed into the car. "This is the first time I've ever been in a police car," I remark.

They burst into laughter, shaking my hand, and ask, "What kind of music would you like to listen to?" Outside the conference centre, onlookers crane their necks to see who is being whisked away in a police car. They must think I'm a notorious criminal, perhaps guilty of stealing food from the buffet. Nonetheless, there's an exhilarating rush as we speed through the city with sirens blaring. Traffic miraculously parts to make way for the police car, allowing us to navigate through the congested streets.

My journey to the airport continues to be a nightmare. The driver remains stuck in traffic due to the relentless downpour, which has only intensified. I've lost sight of Mwangi, so I take matters into my own hands and hail a taxi. We set off, but our progress is immediately halted as we turn the first corner and find ourselves trapped in gridlock. Resignation sets in as I come to terms with the possibility of missing my flight. The anxiety builds within me as the rain pounds on the windshield, and the car windows fog up. Suddenly, the driver's phone rings – it's Mwangi. He has finally arrived at the hotel and insists that we turn back. Frustrated, I make it clear that I refuse to go back as I have a flight to catch. After a heated discussion, we come to an agreement: the driver will not only drop me off at the airport terminus but will also escort me into departures.

Nothing reveals the vulnerability of a city quite like torrential rain, and Nairobi is no exception. We inch forward through the flooded streets, the dirty waters overflowing from drains, and the sewers bursting at the seams. Impassable roads and broken umbrellas scattered in the drains paint a picture of chaos. Each small gain in progress is quickly followed by another halt. Mud covers everything in sight. Finally, against all odds, we arrive at the airport just in the nick of time before the check-in counter closes.

29. GOULASH IN BUDAPEST

**London Heathrow Airport –
Budapest Ferenc Liszt International Airport, Hungary**

Flight time: Two hours and twenty-five minutes

6 April 2016

Arriving in Budapest sparks a flood of memories from my previous visits to the city. This being my fifth time here, I can't help but notice the remarkable changes that have taken place over the years. My first encounter with Budapest was in 1994, just five years after the fall of communism.

Back then, the city wore a dishevelled appearance, with dirty black stone buildings crumbling on the streets. The air was filled with exhaust fumes from decades of benzene, leaving its mark on the city. I stayed in an old communist hotel in the diplomatic quarter, where everything – furniture, decor – seemed to be in shades of dark brown and beige, the colours favoured by the communist party. Cash points were unheard of, shops had empty window displays, and finding a good restaurant was a challenge.

Speaking of restaurants, they often presented an olfactory challenge. As soon as I stepped through their doors, a rancid smell of lard would assault my senses. On one occasion, the stench was so overpowering that I had to leave a restaurant in fear of throwing up right at the entrance. The food, when served, seemed to come out of tins or packets, reminding me of my childhood meals. I vividly remember the chef's attempt at boil-in-the-bag smoked mackerel, which turned out to be the breaking point. It was worse than a lardy croissant I had for breakfast. When I accidentally dropped it, the impact left a mark on the floorboards, as if a grenade had exploded.

As a young woman, I encountered a few memorable experiences in Budapest. First, the inhospitable waiters who refused to serve me when I entered cafés alone. They would stand around in packs, staring but rarely taking my order for a simple coffee. Apparently, it was not the norm for women to visit cafés unaccompanied, as these establishments were seen as male-dominated spaces. My confident presence must have thrown them off.

On a brighter note, I fondly recall dining with the president's daughter in a grand dining room of an ornate government building. She was charming and made sure to include me in the conversation, recognising my status as the only other female diner. It was a boost to my self-confidence at the time. Additionally, during one of my visits, Madonna was in town filming *Evita*, a movie based on the life of Eva Perón. Although my attempt to be an extra on the set failed, witnessing the city transform itself into a film set with fake tanks, military Jeeps, and armed soldiers parading through the streets was a sight to behold.

Over time, I watched Budapest gradually transition into a Western European city. The neo-gothic buildings were restored, their facades stripped back to reveal the original stone and former glory. Western high-street shops made their way into the city, and cash points became a common sight on every corner. Restaurants underwent a transformation, offering diverse cuisines from around the world, and the once-hated lard was replaced by olive oil. Although you can still find *salo*, the Slavic word for lard, if you so desire! It remains a delicacy in Hungary and Ukraine.

During my penultimate visit in 2013, accompanied by my mother, we stayed in a well-known four-star hotel in the historical quarter of Buda. The hotel boasted sweeping views of the city, including Buda Castle and the Danube River. We spent our days exploring hidden corners, stumbling upon antique ruins and medieval buildings that reflected the city's diverse past. Of course, we also indulged in the famous spas Budapest is renowned for and had a slightly tipsy night savouring Hungarian champagne. It was on that evening that we accidentally left a restaurant without paying, but they managed to catch up with us eventually.

And now, back in the present, I swiftly drop off my bags at the hotel and join the conference alongside participants from various countries. The primary objective of this visit is to network and promote financial expertise amongst peers. Accompanying me is Martin, a colleague experiencing Budapest for the first time. He's thrilled to have me as his guide, and after a long day of business, we set off to explore the city.

Our journey begins on Andrássy Avenue, where we board Europe's oldest electric metro network. The yellow train rattles into the bustling platform of Oktogon Station, and we disembark at a beautifully tiled station before the Klaxon-like noise signals the closing of the doors. This is the famous 'M1', the yellow line. We retrace our steps, crossing the Chain Bridge that spans the Danube River, and ascend through the colourful and ornate streets that lead to Buda Castle and its grand squares. From the castle, we take in the view of Pest, the newer part of the city, and admire the Parliament building, inspired by the neo-gothic style of the English Parliament. Our exploration builds up an appetite, leading us to decide that we must try goulash, Hungary's national dish. It's a flavourful combination of seasoned beef with an abundance of paprika that truly blows our socks off. After satisfying our palates and quenching our thirst, we stumble our way back down to the river for a final glass of wine, enjoying the vibrant atmosphere that Budapest offers.

30. PANIC IN ISLAMABAD

**London Heathrow Airport –
Hamed International Airport, Doha, Qatar –
Benazir Bhutto (old) International Airport, Pakistan**

Flight time: Twelve hours and ten minutes

25 April 2016

Here I am again, finding myself not thinking through the landing time of my flight. Touching down in Islamabad at 4am, I realise that as a woman travelling alone to a country with a history of terrorism and violence, I should have planned my arrival time more meticulously. Nevertheless, I'm filled with both excitement and apprehension as I step foot in Pakistan for the first time, especially in the darkness of an unfamiliar land, which adds to the potential risk.

Upon arriving at the airport, I notice that it is swarmed with men wearing long beige shirts and loose trousers, making me stand out even more with my vibrant outfit. The hotel driver greets me, and together we navigate through a poorly lit car park to find his vehicle, only to discover it blocked by another car parked horizontally behind it. In the dimness, I can see groups of men huddled around, staring at us as the driver seeks assistance.

There is evidence to suggest that women often feel fear in public spaces, particularly in unlit areas like this car park. I'm aware of the vulnerability and unease creeping in as I wait anxiously, the car doors locked, continuously checking for any approaching figures. Suddenly, out of nowhere, the driver returns with a forklift truck, which lifts the obstructing car high into the air and out of sight. What a relief!

As we continue our journey through the night, we encounter another unexpected event. The car comes to an abrupt halt as soldiers holding machine guns surround us, peering through the windows to inspect the driver and his passenger. We proceed, only to be stopped again and subjected to the same routine. Finally, we arrive at our destination, a heavily fortified hotel, where additional armed security guards patrol the perimeter and rooftops. At least I can breathe a sigh of relief knowing that I'm in a safe place.

With little time for rest, I manage a quick nap and grab a bite to eat before meeting my colleagues to discuss the day's proceedings. Shahida a British Pakistani, and Steve, representing our organisation, join me. We do some last-minute fine-tuning of his speech and wish him good luck before he takes the podium.

The conference kicks off with the ceremonial lighting of a lamp, setting a positive tone for the day's events. Steve's speech is well-received, and I follow him on the podium to promote a recent research publication. To my pleasant surprise, there is a good representation of women leaders presenting today, defying the stereotypes of a patriarchal society. It is worth noting that Pakistan ranks second lowest in the world for gender equality, according to the World Economic Forum's Global Gender Gap 2022 report.[12] Gender equality remains a challenge in the Middle East and countries like Pakistan and India.

As the business portion of the day comes to an end, we have a couple of hours to spare for sightseeing. Built in the 1960s, Islamabad, the capital and seat of government, is a relatively new city. It exudes a modern charm, with well-maintained wide boulevards that resemble those of European cities like Paris or Bordeaux. The city is clean and green, with expensive individually styled houses owned by diplomats and politicians lining the boulevards. It's not what I expected from Pakistan, as the images I had seen usually depicted cities like Karachi or Lahore, sprawling metropolises often troubled by various social and economic issues.

Stepping out of the car, the sizzling heat hits us as we visit the Faisal Mosque, reportedly the largest mosque in Pakistan. It held the title of the largest in the world until mosques in Saudi Arabia surpassed its size, making it now the sixth largest. Named after the late King Abdul'Aziz of Saudi Arabia, who supported its development in 1966, the mosque stands as an impressive eight-sided concrete shell on an industrial scale. As we stroll around, I attract some unwanted attention from men and boys wanting to have their photos taken with me. Shahida finds the situation quite amusing, while Steve keeps his distance.

Next on our itinerary is a visit to the Centaurus Mall for a shopping spree. Shahida, a self-proclaimed shopaholic,

---

12   https://www3.weforum.org/docs/WEF_GGGR_2022.pdf

finds herself right at home amidst the female population of Islamabad, who share her passion. Clothes are relatively affordable here, but the quality varies greatly, and items disappear from the shelves in a flash. Racks of *shalwar kameez*, the traditional long shirts with trousers, vanish before our eyes. Coat hangers hang naked on the rails as new designer clothes are snatched up within minutes. I make a dash to grab an outfit that catches my eye.

Apart from the presence of heavily armed police and soldiers inside and outside the shopping mall, everything seems relatively normal. In the short time I've been in Pakistan, I'm becoming accustomed to the high level of security and surveillance.

Upon returning to the hotel, we are joined by Gordon, a Swiss colleague. He shares that his embassy has advised against visiting the Monal restaurant in the Margalla Hills for an evening meal due to the risk of a terrorist attack. The restaurant has only one entrance and exit road, making rescue efforts difficult if something were to happen. Undeterred by this new information, and perhaps a bit foolishly, we decide to proceed without Gordon.

The journey to the Monal is an adrenaline-inducing experience as we navigate the winding rocky road, narrowly avoiding oncoming vehicles that swing around corners at high speeds onto our side of the road. We repeatedly swerve onto the verge, narrowly missing boulders. However, the astonishing view from Monal, the cooler temperature, and the absence of mosquitoes at dusk make it all worth it. We find ourselves surrounded by armed guards holding Kalashnikovs on all sides, above and below the restaurant.

As we enjoy a feast of spicy barbecued food, Shahida's partner, Graham, who is working on a project in Islamabad, joins us. The evening is filled with great conversation about our individual travel experiences. Our plates are piled high with a delicious mixture of curries and salads. Steve is trying a bit of everything and I wonder whether I should caution him about the potential consequences on the lining of his stomach. It's always a risk when travelling, but Steve seems to be throwing caution to the wind.

26 April 2016

It's unfortunate that Steve has fallen victim to Delhi belly, leaving Shahida and me to navigate the city and attend to our business without the need for his constant hand-holding. Our first stop is at the UK Department of International Development, but to our dismay we arrive only to discover that they have failed to arrange a security pass for us. We find ourselves stuck outside the gated community, conducting the meeting on a mobile phone from the back of a taxi. The frustration mounts as we realise we can't even convince them to meet us for a coffee at our hotel due to alleged security risks. Shahida voices her concerns, questioning their focus on our security while leaving us vulnerable like sitting ducks. It becomes evident that they may not have been genuinely interested in meeting us in the first place. Despite the rocky start to the day, we decide to have an early lunch, hoping for a more productive afternoon ahead.

27 April 2016

With Steve feeling slightly better and Shahida embarking on a trip to Karachi, I find myself with a free morning before my flight home. Embracing my inner tourist, I decide to visit the Lok Virsa Museum, a fascinating heritage museum that showcases the history and culture of Islamabad. Known as the first state museum of ethnology in Pakistan, it houses a diverse collection of artifacts that offer insights into the country's rich cultural heritage.

After immersing myself in the exhibits and learning more about the local traditions, I make my way to Benazir Bhutto Airport for my departure. As I bid farewell to Islamabad, I reflect on the experiences, challenges, and unique moments I've had during my time in Pakistan. It has been a trip filled with unexpected twists and turns, but one that has allowed me to gain a deeper understanding of the country and its people.

31. MALAYSIA – SET THOSE BIRDS FREE

**London Heathrow Airport –
Kuala Lumpur International Airport, Malaysia**

Flight time: Thirteen hours

19 May 2016

Jet lag is a constant battle, and my arrival in Kuala Lumpur is no exception. The journey from the airport to the hotel feels like an eternity as I find myself trapped in yet another frustrating traffic jam. It seems to be a universal experience in big cities.

Kuala Lumpur, like many other bustling metropolises, boasts a skyline adorned with towering glass buildings, a symbol of its rapid development. However, amidst the modernity, traces of the city's ancient past can still be glimpsed. With a population of nearly 2 million, it's no wonder that Kuala Lumpur is one of the fastest-growing cities in Southeast Asia.

Upon reaching the hotel, fatigue takes over and I head straight to bed. Rest is essential to ensure I am ready for the business engagements that await me in the coming days. The hotel itself is rather unremarkable, just another corporate establishment, albeit with the added convenience of a sprawling shopping mall attached to it. In the sweltering heat and humidity, the mall becomes a refuge, offering a welcome escape from the outdoors.

As my stay in Kuala Lumpur progresses, I find myself traversing the mall multiple times, exploring its vast corridors and seeking solace in its air-conditioned embrace. It's not just about shopping, but also about finding moments of respite and rejuvenation amidst the demands of my busy schedule.

20 May 2016

This morning, we embark on our training event focused on anti-corruption with a group of government accountants. While Malaysia is considered to have lower levels of corruption compared to some other Southeast Asian countries, concerns persist regarding alleged corrupt practices amongst public officials. The participants initially appear reserved, but as the session progresses, they become more engaged, leading to a lively and candid discussion. Our goal is to provide valuable insights and foster an atmosphere of open dialogue.

Later in the day, we proceed to a series of meetings with accounting professionals. Given the deep divisions in Malaysia related to race, religion, and reform, we exercise caution in steering conversations away from potentially sensitive political topics. The country has a complex history shaped by these issues, with the Malay majority often enjoying protected status while ethnic minorities have faced challenges and disparities in their rights and societal standing. These dynamics have contributed to various forms of polarisation and political shifts over the years. In light of these complexities, we choose to keep our interactions light-hearted and engage in topics such as the weather to maintain a harmonious atmosphere.

Tonight, as guests of the training program organisers, we continue to foster positive connections with our hosts. It is important to create an environment where meaningful exchanges can take place, even if we choose to navigate away from contentious subjects. By focusing on building rapport and promoting understanding, we hope to contribute to a more inclusive and collaborative future for Malaysia.

21 May 2016

The day continues with a series of meetings that resemble another mini 'Olympics' of accountants, where professionals gather to network, exchange ideas, and share their latest research. However, it becomes apparent that the meetings are prolonged by a few individuals who seem to enjoy the sound of their own voices, causing the discussions to drag on longer than necessary. Despite this, valuable insights and connections are made throughout the day.

In the evening, we have the pleasure of wining and dining in the presence of the magnificent Petronas Towers, a true symbol of pride for Malaysia and currently the tallest twin towers in the world. As I gaze up at their awe-inspiring height, a sense of exhilaration washes over me, although the alcohol we indulge in may also contribute to my giddiness. With the evening filled with laughter and conversation, it is likely that some of us will wake up with sore heads in the morning, a small price to pay for a memorable night in such a remarkable setting.

22 May 2016

On my birthday, I find myself being woken up at the crack of dawn by enthusiastic colleagues eager to visit the Batu Caves, a temple complex located just outside the city. Although I appreciate their excitement, I have other plans in mind for the day and would rather not venture out so early.

Instead, I decide to immerse myself in the vibrant streets of Kuala Lumpur, traversing between the modern urban landscape and the echoes of an ancient past. After a short journey on the metro, I arrive at Bandar station, a magnificent structure reminiscent of the Taj Mahal with its stunning white domes and golden spires. From there, I make my way to the Islamic Arts Museum, a treasure trove of exquisite artifacts and artworks from Persia and the Middle East. The museum's turquoise-coloured domes add a touch of splendour to the city's skyline. I take a moment to relax amidst the beautifully tiled water troughs in the museum's grounds, enjoying the peaceful ambiance before succumbing to the allure of the museum shop.

With my newfound treasures in hand, I continue my explorations and head to the Bird Park. Although I have reservations about visiting birds in captivity, my Malaysian colleagues assure me that the park promotes free flight and natural breeding. While some of the birds do have the opportunity to roam more freely under large nets, others, like birds of prey, are confined to spacious cages. While I appreciate the opportunity to observe peacocks, hornbills, and numerous other bird species up close, a part of me still longs to see them in their natural habitats.

After an eventful and thought-provoking day, I make my way back to Bandar station and decide to wind down at the hotel's rooftop bar. With a glass of sparkling beverage in hand, I enjoy the mesmerising sunset as the day draws to a close, reflecting on the experiences and memories I have gathered throughout my time in Kuala Lumpur. Soon, it will be time to prepare for my flight back home, bringing an end to this birthday adventure.

32. LIVING IT UP IN PARIS

St Pancras Station, London –
Gare Du Nord, Paris, France

Travel time: Two hours and twenty-nine minutes

10 July 2016

Paris, the city of love, holds its charm even when I visit for work purposes, much like my regular trips to Brussels. As I step out of Gare du Nord, one of the bustling train stations of Paris, I cross the wide boulevards and flag down a taxi to head towards the Organisation for Economic Co-operation and Development (OECD) located at the *Château* de la Muette.

Today, I have a meeting with officials from the ministries of finance of member countries to discuss research and new developments in accounting policies. I am eager to gain insights into how governments are adapting their finance functions to address the financial challenges of our time and, of course, to network with colleagues from around the world.

The meeting starts off well, but with more than 70 participants representing different countries, it becomes a challenge for the chairperson to ensure equal participation while managing the limited time, especially with the language barriers. As the afternoon progresses, fatigue sets in, and participants start losing focus. Headphones for translation are left abandoned on the desks, and some attendees are seen browsing the internet for any travel updates. However, as the day comes to an end, I feel satisfied that I have accomplished my networking goals and gained valuable insights into international accounting developments.

As we conclude the proceedings, we are directed into the magnificent eighteenth-century building of *Château de la Muette*. Surrounded by lush woodlands, meticulously maintained gardens, and prestigious Parisian properties, it provides a perfect setting to unwind after a long day. I cannot recall a time when we weren't treated to delicate hors d'oeuvres served with a glass of sparkling wine, adding a touch of elegance to our post-meeting experience.

33. HOT IN HYDERABAD

London Heathrow Airport –
Zayed International Airport, Abu Dhabi –
Rajiv Gandhi International Airport, Hyderabad, India

Flight time: Thirteen hours and ten minutes

19 October 2016

It's 3am in Abu Dhabi Zayed International Airport, and I find myself wide awake and caffeinated, drafting emails and hitting the send button in a monotonous routine. The soulless airport atmosphere makes me feel bored and disconnected from any specific location. As I click send, the cleaners rush around, vacuuming the area in preparation for the early morning travellers.

I eagerly keep an eye on the flight information screen, waiting for the Hyderabad departure time to appear. There's a hint of nervousness in me as I hadn't had the chance to apply for a business visa before leaving London, and I'm relying on a tourist visa instead. To ease my worries, I utilise my time in the airport to plan for any possible questions that the border control officer might ask upon my arrival. I commit to memory the top ten tourist sites, including an ancient fort and the iconic monument Charminar with its four minarets.

The sweltering heat of Hyderabad hits me as I stand in the queue at border control, and I can feel hot flushes creeping in. The border-control officer scrutinises my passport security before asking the dreaded question, "Why have you come to Hyderabad?"

I confidently reply, "I'm here for a sightseeing visit, and Hyderabad has been on my bucket list for a long time." The officer then enquires about the sights I'm hoping to see. Thanks to my early morning planning in Abu Dhabi, I am prepared with a detailed response, and by the time I reach the third tourist attraction, the officer seems to lose interest.

Passing through security, I notice Sam, the organiser of this business visit, stuck at border control. He wears a golf cap and appears hot and flustered. I assume he is also on a tourist visa but took a different approach, pretending to be here for golf. I chuckle to myself and continue on, knowing that our varying tactics have helped us navigate the entry process.

With the remaining few hours of the day, I collapse into bed, succumbing to the familiar grip of jet lag.

In the evening, we gather at the 'waterfront restaurant' situated along the shores of Hussain Sagar Lake in the heart of Hyderabad. We are in high spirits, celebrating the lifetime contributions of a colleague to the accountancy profession. Although the service is a bit slow, the moment the food arrives, I'm captivated by the flavours of the honey chili potatoes and dal. They are so delicious that I selfishly keep them to myself, savouring every bite.

As I glance out the window, my gaze falls upon a giant Buddha statue standing on a rock, commanding attention over the lake. However, alongside its imposing presence a foul smell wafts in, souring the air and clinging to my skin and clothes. I'm informed that the smell is caused by floating waste, including polythene bags, plastics, sewage, and industrial effluents. Despite past clean-up efforts, little has changed and the situation can only improve with a concerted effort to address the source of the pollution. It saddens me to witness the deterioration of this natural beauty due to neglect and lack of proper waste management.

20 October 2016
The wait for the Minister of Finance to arrive at the conference feels like an eternity, and I find myself twiddling my thumbs alongside other participants who are growing restless. Finally, the conference organisers decide to start the proceedings with the first panel of speakers. However, as soon as they take the stage, they are abruptly removed as the finance minister enters the auditorium. The sudden change disrupts the flow, and one disgruntled speaker decides to leave in frustration and catch a flight home. The opening ceremony commences with the lighting of the lamp, symbolising brightness and prosperity for the conference.

When lunchtime arrives, there is no food in sight, as a compulsory photo session prolongs the wait. Group photos, individual photos, various group combinations, and gender-sensitive photos seem to go on forever. It's an annoyance that always seems to take place before lunch or at the end of the day when hunger sets in. The photoshoot is a constant source of frustration.

In the mid-afternoon, my colleague Shireen and I take a taxi to the Golconda Fort, located just outside the city. The fort's name translates to 'Shepherds Hill', originating from a mythical story of a shepherd boy who discovered a god idol on the hill. Intrigued by the boy's discovery, the ruler of the Kakatiya dynasty decided to build a fort on the hill as a place of worship.

The fort is a remarkable stone structure dating back to the eleventh century. It presents a complex maze of intricate carvings, mosques, and pavilions, most of which now lie in ruins. The acoustics within the fort are impressive, as our voices bounce off the walls and reverberate through the air. We explore the nooks and crannies of the fort, seeking shelter from the scorching sun and intense heat. Climbing to the top mosque seems unbearable given the extreme temperatures.

Afterwards, we hop back into the taxi and make our way to the Charminar Bazaar market, known for its vibrant and rustic charm. As we enter the bustling streets, we encounter a sea of people going about their daily lives. Women adorned in colourful sarees pass by, catching our attention, while children on bikes curiously stare, likely intrigued by my blonde hair. The market's density leaves no open space and the energy and liveliness are palpable.

We navigate the narrow streets of the market, weaving through the crowds, and come across shops selling a variety of bangles, handwoven fabrics made of silk, cotton, brocade, and gold embroidery. The air carries a fragrant blend of bergamot, jasmine, and incense, with undertones of rose, patchouli, and vanilla. It's an intoxicating and sensory experience unlike any other market I've visited.

Shireen and I reside in London and have crossed paths in various parts of the world, but never managed to meet in our own city. This pattern has persisted for almost a decade, and it seems unlikely to change anytime soon. Amongst Shireen's admirable qualities is her dedication to community work, particularly supporting refugees fleeing war-torn countries. Today, she is on a mission to purchase materials for a refugee project she is involved with in Greece. These materials will be used by women refugees to create clothes that they can sell to support their families and find purpose in their lives. Our task is to find high-quality materials at bargain prices. Shireen's bargaining skills prove fruitful, and I quickly learn that negotiation is essential to avoid being seen as an easy target for shopkeepers looking to hike their prices. With one successful purchase behind us, we continue our quest to find shocking pink silk for my own wardrobe.

Despite the crowded streets, we repeatedly encounter male colleagues during our shopping excursion, much to their amusement. Many of them carry shopping bags filled to the brim with sarees and materials to take home to their wives. It's an act that requires bravery, as getting it wrong could have its consequences!

As I prepare for my late-night flight back home, I feel less concerned about border control officers questioning my visa status. I breeze through the airport, take my seat on the plane, and eagerly gulp down a glass of champagne, savouring the journey ahead.

34. SMALL PLEASURES IN HONG KONG

**London Heathrow Airport –
Hong Kong International Airport, Hong Kong**

Flight time: Twelve hours and forty minutes

19 November 2016

Excitement fills me as I embark on my long-awaited trip to Hong Kong, a vibrant administrative region of the People's Republic of China since 1997. While I have played host to visitors from Hong Kong in London, the tables have now turned, and I have the opportunity to be their guest. My purpose for the visit is to attend an academic conference at City University of Hong Kong located in Kowloon.

The prospect of exploring one of the world's financial and trade centres against the backdrop of the South China Sea fills me with anticipation. Upon arrival, I hop into a taxi at the airport's exit hall and head towards Kowloon in the northern part of Hong Kong, where I will be staying for the next few days. The journey is accompanied by the dazzling glare of headlights from the never-ending flow of oncoming traffic. As we approach the city, a mesmerising cluster of illuminated skyscrapers emerges, seemingly rising out of nowhere. Hong Kong's population density is amongst the highest in the world and in Kowloon alone, where I will be residing, the population reaches 2.1 million. The taxi drops me off in a bustling side street, pointing me in the direction of the hotel. However, amidst the shops selling an array of cheap goods, the only visible landmark is a squeezed-in sex shop. Thankfully, a helpful mobile-phone shop owner eventually directs me to the hotel entrance in a dimly lit square.

The hotel itself proves to be peculiar, adorned with health and safety notices cautioning guests about typhoon preparedness. One notice stands out, emphasising that if the eye of a typhoon passes directly over Hong Kong, there may be a temporary lull lasting a few minutes or hours. It warns against relaxing one's guard, as violent winds from a different direction will soon resume. The reminder to remain in a protected area and be prepared for destructive winds is duly noted. After depositing my luggage, I venture around the corner to a local restaurant, where I find my colleagues already struggling with their chopsticks, providing some amusing moments.

I am eager to immerse myself in the vibrant atmosphere of Hong Kong and make the most of my time at the conference, while also exploring the city's diverse cultural and culinary offerings.

20 November 2016

As I make my way through the concrete jungle surrounding the university, I can't help but notice the distinctive brutalist architecture that defines the campus. The buildings are constructed primarily of cement, exuding a solid and imposing presence. This style of architecture, known for its utilitarian aesthetic, has garnered various architectural accolades. The long walkways are made of the same grey material, creating a monotonous visual landscape. Despite its unconventional beauty, the university hall buzzes with activity, filled with professors and students hurrying between seminars and lectures.

As a practitioner in the field of accounting, I have the privilege of chairing one of the later panels. I playfully request that the attendees treat me gently, and they respond with polite consideration, ensuring a fruitful and respectful discussion. Throughout the day, navigating the concrete walkways becomes a tiresome task as I move from one topic to another or make my way back and forth to the hotel.

However, the evening brings a welcome reprieve in the form of a dinner and award ceremony. The atmosphere is lively and filled with the cheer of celebration, accompanied by a healthy dose of revelry. Amidst my busy schedule, I also find solace in a small pottery shop that I frequently pass by. The shop offers a delightful selection of ornate pottery at remarkably low prices. I can't resist purchasing six Chinese bowls at the steal of fifty pence each, a fraction of what they would cost in London. To this day, I continue to use and cherish these bowls, serving as a lasting reminder of my time in Hong Kong.

21 November 2016

Today follows a similar routine as yesterday, but I'm eagerly awaiting the post-lunch exploration time that awaits me in the vibrant city of Hong Kong. I embark on my adventure by tramping through the streets of Kowloon, which was once a walled city. The remnants of its colonial influence still linger, and every corner reveals something new and exciting, from colourful markets to quirky florists peddling exotic plants. Discovering these local and untouristy markets is truly one of life's little joys.

I continue my meandering journey down to the waterfront, where the legendary *Star Ferry* awaits. This iconic passenger ferry shuttles people across Victoria Harbour, bridging the gap between Kowloon and Hong Kong Island. As the ferry glides through the harbour, I catch glimpses of the skyscrapers towering above the low-lying fog, giving me a panoramic view of this densely populated city. What makes the *Star Ferry* truly extraordinary is not only its 120-year-old history as the oldest form of public transport but also the captivating sight of other boats passing by. From working barges to sea-fishing vessels, the water comes alive with activity. I even spot fishermen and their wives on deck, meticulously gutting and cleaning fish as they dart across the water.

Stepping off the *Star Ferry*, I find myself amidst a crowd of tourists eager to explore Hong Kong Island. The island buzzes with life, featuring covered walkways suspended above the bustling streets, weaving their way through shopping precincts and office blocks. These covered pedestrian walkways are an absolute godsend, especially when it starts to rain. It's clear that I've entered the financial heart of Hong Kong, with the imposing Shanghai Bank (HSBC) Building and China Bank Tower standing tall nearby, alongside other renowned banks.

As I venture further, I stumble upon another side of Hong Kong Island, with its charming winding streets adorned with intriguing shops, cosy cafés, and enticing restaurants. I find myself on Cat Street, home to a plethora of antique shops brimming with communist relics, including Chairman Mao posters and small pottery figurines depicting people from various walks of life during the Mao era. The street stretches from the Central Police Station to the Man Mo Temple, where I decide to make a quick stop and light a joss stick for reasons unknown to even myself!

Although I hadn't planned on making any purchases, I couldn't resist adding another figurine to my collection of Mao memorabilia. This particular figurine depicts a teacher proudly holding up the red book filled with quotes from Mao's speeches. It reminds me of the time I bought a figurine in Xian, China. The shop owner's enthusiasm leads to an unexpected lunge for a hug. Caught off guard, I instinctively step back, only for him to reassure me that his intentions were purely friendly, as he was too old for any romantic gestures. We shared a good laugh, bid our farewells, and I made my way back through rows of shops offering traditional homewares, Buddha images, and jade accessories, eventually reaching the harbour.

Returning to Kowloon, I make sure to indulge in one last treat before bidding farewell to Hong Kong. The Peninsula Hotel, a beloved institution amongst tourists, beckons me for a delightful tea session. Although I forgo the traditional three-tiered set of cakes, scones, and sandwiches, the sumptuous cake paired with a glass of sparkling wine is simply divine and provides a fitting conclusion to my day's adventures.

22 November 2016

In the midst of my airport routine this morning, a rather peculiar incident unfolds, something that I haven't encountered before. A gentleman from Bahrain, dressed in traditional attire, approaches me at the check-in desk, claiming that he is on the same flight. He appears somewhat nervous but kindly offers to help me with my heavy bags as we navigate through the airport. It's not every day that a stranger offers such assistance, so I'm pleased to accept his offer.

As we make our way, he begins sharing his story of being swindled out of $50,000 in a shady property lease deal, expressing his relief at leaving Hong Kong behind. Despite his anxious demeanour, he becomes slightly irritable when I decline his invitation to join him for breakfast. After hastily finishing his coffee, he leaves me his phone number and disappears, never to be seen again. It leaves me with a slight unease, prompting me to double-check my bags, just in case some illicit substance had been planted within them. He doesn't board the plane!

Since my visit to Hong Kong, both the social and political landscape have deteriorated, and it saddens me to witness the university and walkways being used as barriers against the increasingly stringent measures imposed by Chinese rule. However, the people of Hong Kong have shown remarkable resilience, and it is my fervent hope that they will continue to thrive, making Hong Kong one of the most vibrant and successful places in the world to live and work.

35. MY OLD FRIEND, DHAKA

**London Heathrow Airport –
Hazrat ShahJalal International Airport, Dhaka, Bangladesh**

Flight time: Thirteen hours

29 January 2017

Back in Dhaka, I once again find myself faced with the age-old dilemma of whether to resort to bribery to expedite the visa process upon landing. However, my steadfast principles prevail, and I choose to patiently endure the long queue like everyone else. Obtaining visas can be an arduous affair, demanding an absurd amount of detailed information. I recall the time I had to provide my blood group for a Pakistan visa. With the urgency of needing the visa quickly, I didn't have time to find out my actual blood group. So, in a stroke of resourcefulness, I resorted to Dr. Google and confidently declared myself as blood group O – the most common blood group around.

As I traverse through the city, little appears to have changed since my last visit. The familiar chaos, pollution, and unfinished elevated highways still dominate the landscape. Naturally, I find myself running late for my first business meeting. Manoeuvring through the traffic, our taxi suddenly emits a burst of black smoke, forcing us to a complete halt on the congested highway. The resourceful taxi driver manages to navigate our vehicle to the side of the road and there we wait, seemingly trapped in an eternity, for a replacement taxi to come to our rescue. I finally arrive at my meeting, slightly dishevelled in my jeans and T-shirt, since there was no time to change into my business suit. The day becomes a whirlwind of back-to-back advocacy meetings with government officials, and I am immensely relieved when it finally draws to a close, allowing me to retreat to the hotel for a well-deserved shower.

30 January 2017

Another eventful day begins with a seminar on government accounting, and as is often the case, the room is predominantly filled with men. During the breaks between speeches, I manage to sneak a glance at the newspapers to catch up on the latest headlines. The Dhaka Tribune highlights the accountants' call for greater accountability and transparency in financial institutions, which they argue are jeopardising the country's stability.[13] Amongst other news stories, there's a report of a local paddy field drowning in toxic waste dumped by a food company, posing a threat to hundreds of hectares of land.[14] A particularly grim story involves the hacking death of a schoolboy, highlighting

---

13 "President for ensuring financial transparency", Dhaka Tribune, 29 January 2017

14 "Boro field drowned by toxic waste of ACI", Dhaka Tribune, 29 January 2017

the need for improved security measures.[15] Environmental degradation, from wetland exploitation to hindering river flows through urban development, also takes centre stage. The impacts on farmers and fishermen are severe, with concerns about their gradual extinction.

Although my visit has been brief, with limited opportunities for exploration, one memorable moment stood out. It was when we drove through a vibrant food market. The roadside was adorned with vividly coloured produce displayed on straw mats, while market vendors hurriedly carried baskets brimming with vegetables and fruits on their backs. The air was filled with the enticing aromas of sweet and sour delicacies, evoking the best of Bangladeshi cuisine. The food I had in Dhaka was truly exceptional and the dal, in particular, left me craving for more. It was a delightful culinary experience that also managed to keep the dreaded Delhi belly at bay!

For my final evening, a colleague who runs a business takes me to a fast-food restaurant in the city. We enjoy a satisfying meal of rice and fish curry, accompanied by a bottle of cola. As we chat and savour the evening, a new adventure unfolds when I find myself trusting my life in his hands as we attempt to cross a poorly lit road that resembles nothing more than a rubble-filled dirt track. With relentless traffic zooming past, we reach the central reservation, only for me to stumble over a concrete bollard. Thankfully, the car comes to an abrupt halt, sparing me from a potential hospital visit. It's moments like these that remind me of the unpredictability of travel!

---

15 *"Schoolboy hacked to death"*, Dhaka Tribune, *29 January 2017*

36. THRILLER IN MANILA, CONTINUED

**London Heathrow Airport –
Ninoy Aquino International Airport, Manila, Philippines**

Flight time: Thirteen hours and twenty-five minutes

2 March 2017

It feels surreal to be back in Manila after three long years. The flight itself seemed to drag on forever, and even the notorious Manila traffic jam welcomed me back with open arms. As we crawled along the highway, the familiar sight of pollution clouding the skies and obscuring the sun greeted me. But despite the challenges, I'm here to lend support to a financial management conference and engage in discussions about financial management reform.

Amidst settling in, I receive a WhatsApp message from Keith, a colleague who suggests going out for an evening meal. However, the exhaustion from the fourteen-hour journey leaves me with little energy to spare, so I reluctantly decline the invitation. Little did I know at the time that my decision was a stroke of luck. It turns out that Keith, unfortunately, succumbs to food poisoning and is bedridden for the next three days. It's moments like these that make me appreciate the fortune of having avoided the dreaded Delhi belly during my travels.

3 March 2017

The conference kicks off, and I find myself on the podium alongside several male speakers. It becomes evident that I'm the only woman in the group, and unfortunately, this realisation plays out in subtle ways. The chair tends to direct questions to my male colleagues, even when I know I have better answers. This frustrates me, but an unexpected twist occurs when a fellow panellist keeps passing on the questions for me to answer. The chair grows increasingly agitated, pressing the panellist for a response, but the microphone continues to find its way into my hands. I'm more than capable of handling the questions, but their behaviour makes both the chair and the panellist look rather foolish in front of the audience.

In the early evening, I meet Fred, a local resident of Manila whom a colleague had suggested I meet. He arrives dressed quite stylishly and presents me with a thoughtful gift of dried peach halves and mangoes, a specialty of the Philippines. It's a warm and welcoming gesture that sets a positive tone for our evening.

We venture to a typical Philippine restaurant where we delve into various topics, ranging from politics to sports, and our conversation is filled with laughter. However, as we touch on the political landscape of the Philippines, I become aware of a darker side. Fred shares stories of the shadowy aspects of politics, particularly the alarming toll of the so-called 'war on drugs' initiated by President Rodrigo Duterte. Thousands of lives have been lost, with both state forces and vigilante groups carrying out ruthless killings under the pretext of fighting the drug trade. Fred lowers his voice as he shares these details, cautiously looking around to ensure no one overhears our conversation. As the evening comes to an end, I'm left with a lingering sense that there is more beneath the surface in this country.

Upon returning to the hotel, I pick up a very old copy of the *Khaleej Times* and stumble upon an article about a Philippine drug lord being held in Abu Dhabi.[16] The report highlights the Philippine government's urgent call for his immediate repatriation, as he ranks second on their most-wanted list of individuals involved in the illegal drug trade. It serves as a reminder of the complex issues surrounding the drug trade and the efforts made to address them.

4 March 2017

This morning takes me back to the stunning buildings of the Asian Development Bank for a series of meetings that wrap up just in time for lunch. I meet up with Fred, and we decide to grab a bite to eat before indulging in a little shopping adventure to find some genuine pearl earrings. We start off at a charming café nestled in the heart of a bustling shopping precinct, which instantly reminds me of the delightful dining experiences I had on my previous visit. Energised by our meal, we embark on our quest for the perfect pearls, knowing that Fred will lead us in the right direction.

---

16 *"Philippines Drug lord held in Abu Dhabi"*, Khaleej Times, 18 October 2016

Later, I join my colleagues for a city tour, and we hop aboard a jeepney – a vibrant and flamboyantly decorated vehicle that resembles an elongated military Jeep. These iconic vehicles, remnants of the American occupation during World War II, have become both a national treasure and a notorious presence on the roads. Originally abandoned by the Americans, Filipinos transformed them into public transportation by adding benches and adorning them with colourful illustrations. Today, they outnumber buses significantly, although they also contribute to a significant portion of road-vehicle pollution. The government aims to phase them out, envisioning them zooming off into the sunset.

Our tour takes us past the cathedral and towards Fort Santiago, a historic museum and park constructed as a stone fortress by the Spanish in the seventeenth century. The scars of World War II are still visible on its walls, providing a poignant reminder of the past. At the fort, we delve into the history of the Philippines' independence movement, which sought freedom from Spanish colonial oppression. Notably, the national hero José Rizal, often referred to as the 'George Washington of the Philippines', was imprisoned here prior to his execution during the start of the Philippine Revolution in 1896–1898. We have the opportunity to see the prison cell where he was held and pay homage to him at a shrine that replicates his ancestral home.

Our next destination is Rizal Park and Plaza, where we disembark for a group photo. It was at this very park that José Rizal was unjustly executed in 1896, falsely convicted of leading the rebellion, despite his commitment to nonviolent resistance. His untimely death marked a turning point in the Philippine revolution, galvanising the fight for independence.

Continuing our journey through the congested streets of Manila, we find ourselves surrounded by the thumping beats of acid house music emanating from every corner of the jeepney. Our driver appears a bit disoriented, struggling to locate our desired shopping precinct. As we go round and round in circles, the music grows louder, adding to our frustration and weariness. Eventually, we request the driver to drop us off on a bustling ring road, determined to find the shopping mall on our own. Before long, we stumble upon the entrance to the Venice Grand Canal Mall.

To say that the mall is enormous would be an understatement, but what takes me completely by surprise is the perfect replica of the grand canal in Venice, complete with the iconic Rialto Bridge and even working gondolas. I'm left speechless, pondering the reasons behind recreating Venice in the heart of Manila. We select a restaurant situated by the canal and opt for the local specialty – Italian cuisine. It's undoubtedly one of the most peculiar experiences I've had, once again finding ourselves dining in a shopping mall. As the drinks flow, we can't help but crave a ride on a gondola. Alas, we arrive too late, and the gondolas have already ceased operations for the evening. We hop back onto the jeepney, and as the music resumes, we find ourselves happily singing along, embracing the quirkiness of the moment.

5 March 2017

It's my final day in Manila, and today is a well-deserved break from work. I'm waiting in the lobby for Sam and Ben, two colleagues who will be joining me on an excursion to Tagaytay, south of Manila, where the impressive Taal Volcano is located. Sam and I have known each other for years, and although he tends to talk endlessly about audit, he can be good company. As for Ben, he's a tech entrepreneur going through what seems like a mid-life crisis.

We embark on our journey and the drive to Tagaytay offers amazing views of the expansive lake and the imposing volcano before us. Our excitement builds as we transfer to a small outrigger boat to cross the choppy waters and reach Taal. Little did we know that the Taal Volcano is considered one of the smallest and most active volcanoes in the world, responsible for significant eruptions in the past, claiming numerous lives and leaving destruction in its wake. Our guide assures us that we'll be safe and we take a calculated risk, unaware of the eruption that would occur in 2020, wreaking havoc and covering the surroundings in ash.

Upon reaching the volcano, like typical tourists, we decide to scale it on horseback. It's an adventure fraught with peril. I struggle to maintain my balance on the horse as it ascends, dodging piles of pony poo along the way. Plumes of smoke rise from the ground, and the distinct smell of burning sulphur fills the air. Despite the challenges, I manage to reach the top and stand on the viewing deck, looking down into the bluish-green crater with its orange-coloured rocks and small bubbling pockets emitting smoke. The pervasive sulphur scent permeates my clothing, and the intense heat creates an overwhelming dryness, leaving me desperately thirsty. I take a gulp of water before preparing for the descent, trying my best to avoid a clumsy fall. The guide warns me to lean back, but my final dismount is far from graceful as I snag my leg on the horse's reins, providing a moment of hilarity for my colleagues who capture the incident on video.

As we make our way back to the mainland on the boat, we face strong incoming winds that send waves crashing over us, soaking everyone on board. In the distance, we can hear people cheering on cockerel fights, a popular sport in the Philippines that continues to this day.

And so, my thrilling adventure in Manila comes to an end, leaving me with unforgettable memories and a few amusing mishaps captured on film.

37. PUNCHING THE AIR IN ZIMBABWE

**London Heathrow Airport –
Robert Gabriel Mugabe International Airport, Harare, Zimbabwe**

Flight time: Twelve hours and thirty-five minutes

18 March 2018

We touch down at Mugabe International Airport, and there stands Donald, my highly recommended protector for the next few days. As we approach his white Fiat, I can't help but notice the cracked windscreen and the missing side-door mirror. Let's hope the rest of the car is in working order, or else we might find ourselves stranded along the way.

We embark on the journey, bouncing up and down on roads that resemble obstacle courses with potholes big enough to swallow any unsuspecting tyre. Just as we come to a stop at a junction, a young boy materialises out of thin air, brandishing a long-handled hammer and mallet. My mind races with worries about what he intends to do with those tools, but Donald calmly winds down the window. To my surprise, he reaches into the glove compartment and hands the boy some cash. Apparently, this is no protection racket; the money goes towards the boy's labour in filling in the potholes. These street boys have taken on the job of public servants, stepping in where the government falls short due to dire financial circumstances. The Zimbabwean economy is in free fall, and the people are left to their own devices, forced to rely on their own resourcefulness to tackle everyday problems.

Harare presents itself as a mishmash of unkempt buildings, almost frozen in time, reminiscent of the 1970s. Decades of neglect have left the city's infrastructure in a sorry state. The economic crisis in Zimbabwe, in part a consequence of Mugabe's legacy, lingers like a dark cloud. Mugabe was ousted by the military, and the subsequent elections brought Emmerson Mnangagwa to power, his rallying cry being 'Zimbabwe is open for Business'. The hope was to attract international investment and aid to help alleviate the country's substantial debt. However, the current economic situation has resulted in a severe cash shortage, rendering government bonds practically worthless. Queues form at cash tills, and even when I try to pay for a simple coffee, there's no change available. Taxi drivers, too, struggle to provide any change.

The hotel I'm staying in is an old colonial establishment, boasting beautiful gardens and sculptures – a perfect sanctuary for unwinding after a long day of business. This evening takes an intriguing turn as I inadvertently eavesdrop on a conversation between two guys from different international aid organisations, meeting after a considerable hiatus. Harare is teeming with staff from various donor and aid organisations, sometimes interacting, sometimes keeping to themselves.

Their conversation commences with one guy expressing his limited time, insisting on having just one beer. Yet, as if caught up in a game of one-upmanship, they both try to outdo each other, boasting about the size of their respective teams and the complexity of their capacity-building projects. It doesn't take long for the guy who claimed he had little time to be convinced to stay for a second beer. The competition intensifies as they spar over who is leading the most strategically important project in the region. Another round is swiftly ordered, their voices growing louder as they reach a higher gear. Now they delve into dangerous territory, recounting stories of projects marred by threats of violence and kidnapping. Of course, they both have tales to share! Like birds preening their feathers, one outshines the other, and by the time they reach their fourth and fifth beers, they are convinced they are running the world. It's been a night of entertainment, and I find myself unable to suppress the smiles that keep creeping onto my face.

19 March 2017

This morning, I'm in for a surprise as Donald is a no-show. Instead, a toothless pensioner wearing a flamboyant and vibrant shirt steps in as my chauffeur. I brace myself, holding onto my seat tightly as he zooms through town, only to abruptly halt in the midst of a traffic jam. But I quickly realise that I'm in capable hands as this seasoned driver knows all the secret shortcuts. We navigate through minor roads, zigzagging until we reach a junction leading to a major arterial road. With unwavering confidence, he ventures forth, daringly crossing all four lanes and bringing the oncoming traffic to a standstill. As we successfully make it to the other side, he triumphantly punches the air, celebrating his feat of halting Harare's morning commuter traffic. I've never experienced anything quite like it, and I can't help but have a blast on my way to work. Farewell, Donald, for now, this charismatic driver has won me over as my preferred companion on the road.

Today's agenda involves a gathering of approximately two hundred government officials from both central and local governments. All are united in their aim to enhance government financial management in the country. The expectations are sky-high, matched only by the collective commitment to tackling the financial challenges at hand. As I learn more about the outdated financial systems and pervasive corruption, I realise the steep uphill climb they face in achieving meaningful improvements. Nevertheless, this assembly of experienced individuals marks an important initial step toward addressing these pressing issues. After a long but enlightening day, it's finally time to retreat to the comforts of the hotel and indulge in some well-deserved time for leisure.

As I enter the bar this evening, there are no international development aid workers to eavesdrop on. Instead, I find solace in a local shop brimming with enticing tourist goodies. The handbags catch my eye, but I have a feeling that strolling through Crouch End with a zebra-skin bag might attract the wrath of animal-rights activists. So, I opt for locally embroidered linen, a tasteful and culturally appropriate choice. Who knew that shopping could be such a delightful adventure?

20 March 2017

As the morning begins, we gather for a final wash-up session, recapping the insights and achievements from yesterday's fruitful discussions. Together, we design a plan to move forward, ensuring that the momentum generated during our time together doesn't fade away. Finally, as I receive the green light, I bid farewell to my colleagues and eagerly anticipate the leisurely afternoon that awaits me by the inviting swimming pool.

The poolside is adorned with captivating sculptures of ducks and wild cats, lending an air of whimsy to the serene setting. I plunge into the cool water, relishing the soothing respite from the scorching afternoon heat. With each stroke, I reflect upon the remarkable experiences and connections I've made during my stay in Harare. It has been a journey of growth, learning, and forging meaningful relationships.

38. A HIGH RED BLOOD CELL COUNT IN NEPAL

London Heathrow Airport –
Atatürk International Airport, Istanbul, Turkey –
Tribhuvan International Airport, Kathmandu, Nepal

Flight time: Eleven hours
Previous visit in 2015 – return journey:
Forty-eight hours

23 May 2018

As I embark on my third business trip to Kathmandu, I can't help but reminisce about my previous visits. The first was marked by a public strike that threatened to escalate into something more sinister, and the second, just weeks later, narrowly avoided the devastating earthquake that shook the city in 2015. The memory of the earthquake's aftermath, with its devastating impact on the lives of thousands, still lingers in my mind.

Curiosity piques as I wonder how things may have changed since my last visit. As I make my way to the hotel, I reflect on the UK government's threat to cut financial aid due to concerns over the Nepalese government's response to endemic corruption. This issue was extensively covered in the *Kathmandu Post*, highlighting the importance of addressing this challenge.[17] I also recall an intriguing story about the capture of a ringleader involved in the illegal chiru wool trade.[18] Chiru wool, derived from the rare Tibetan antelope, had been banned from commercial trade since 1974. Additionally, I recall news focusing on the government's efforts to preserve traditional houses for their cultural and historical significance, as well as advocating for women's rights and equality in the Constitution.

Amongst these memories, there are also moments of fondness. I developed a taste for Momo's, those delightful steamed dumplings filled with either meat or vegetables, which seemed to effortlessly melt in my mouth. Wandering through the labyrinthine streets, bustling with people and street dogs, and engaging in lively haggling sessions with the local shopkeepers remain cherished experiences.

However, there were also fewer endearing moments, such as finding myself locked outside the airport at dawn, surrounded by packs of scavenging stray dogs. Once I gained entry, the day took a turn for the worse as low-lying fog disrupted flight schedules. The chaos at the airport resembled the bustling streets of Naples, with flights being cancelled, rerouted, or disappearing from the boards altogether. Amidst the overcrowding and confusion, dwindling supplies of food and water became a pressing concern. Yet, there was a silver lining – the abundance of pashminas available for purchase.

---

17 "*Nepal risks UK aid cut for endemic corruption*", Katmandu Post, 28 March 2015

18 "*Ringleader of chiru wool smugglers nabbed*", The Himalayan Times, 28 March 2015

A valuable lesson learnt from previous experiences is the need to surrender to the whims of the airport system once you've checked in. Helpless and without recourse, all you can do is remain patient and go with the flow. Waiting for flight details to appear on the screen took a gruelling twelve hours, and even at the gate, airport staff seemed clueless until a passenger spotted the plane touching down in the distance. The return journey to the UK lasted forty-eight hours, a relatively swift ordeal compared to some of my colleagues' experiences. Let's hope for smoother travels this time around, free from such trials and tribulations.

24 May 2018

The bustling hotel lobby is a sight to behold, with mountaineers bustling about, readying themselves for their Everest conquests and other majestic peaks. Holdalls and climbing gear are strewn across the floor, a colourful chaos orchestrated by the sherpa guides. It's hard to miss the seasoned mountaineers who have returned from their expeditions, sporting dishevelled beards, bulging muscles, and weathered faces that bear witness to the harsh conditions they've endured. You can often find them in the hotel bar, regaling one another with tales of near-death experiences or engaging in friendly competition to prove who has taken the greatest risks.

Since our business program doesn't start until the afternoon, I join my colleagues on a trip to Swayambhu, affectionately known as the monkey temple. Climbing 365 stone steps in the sweltering heat tests my endurance. I'm reminded that getting accustomed to climbing steps is a prerequisite for trekking in the Himalayan foothills. With my heart pounding and sweat trickling down my back, we finally reach the temple's summit, where we're rewarded with heart-stopping views of the sprawling city below.

On our descent, we navigate our way past the snoozing dogs, observing the monkeys grooming each other and skilfully avoiding the persistent peddlers attempting to sell us an array of singing bowls. These traditional meditation tools, usually made from brass or metal, create soothing sounds when struck or rubbed with a mallet. They may seem tempting to purchase at the time but like many trinkets, they often end up as unnecessary clutter once we return home. Some claim that singing bowls can even alleviate joint pain, though the mechanics behind this remain a mystery to me. Nevertheless, my aching leg could use some relief.

The afternoon flies by as we convene to prepare for the conference scheduled for the following day. The Nepalese government is eager to showcase its progress in financial management to the gathering of international donor bodies in Kathmandu, so we approach our preparations with the utmost seriousness. The evening remains subdued as most of us are still grappling with jet lag and seize the opportunity to fine-tune our presentations for the upcoming day.

25 May 2018

The customary lighting of the lamp commences, symbolising the positive and productive discussions to come. The Minister of Finance takes the podium for the inaugural speech, a predictable occurrence given the predominantly male composition of finance ministers worldwide. It's disheartening to note that only 16% of ministers of finance are female, highlighting the ongoing need for progress and gender equality in this field. Participants hailing from different corners of the globe, including supranational donor bodies, are eager to witness the outcomes of their investments and the improvements made. Their shared objective is clear: to channel investment funds into developing countries like Nepal, facilitating their transition from low-income to middle-income nations and ultimately reducing poverty.

Once again, the line-up of conference speakers tilts predominantly towards men, with Shireen and myself standing as rare exceptions. It's frustrating to witness the persistent underrepresentation of women in the accounting profession within governments worldwide. This norm, though widely accepted, is an obstacle that should have been overcome long ago, especially in the twenty-first century. Nonetheless, my efforts in presenting on the intricacies of financial management yield a modest reward – a lovely pashmina to join my ever-growing collection of scarves.

26 May 2018

Today brings a slightly calmer atmosphere as we delve into evaluating the conference's success and shaping future agendas. Opinions abound, but I can't help but wish that the discussions moved beyond the enjoyment of lunch and the adequacy of tea breaks. While it's business as usual for many, my mind is already racing ahead to tomorrow's trek, brimming with anticipation and a hint of trepidation.

During teatime, Sam and I rendezvous with our sherpa guides for a safety briefing, eager to uncover what awaits us on this expedition. Somehow, I sense that this trek won't be a leisurely stroll in the park. My excitement builds!

27 May 2018

The early morning journey from the city is fraught with traffic, making our progress to Sundarijal slow. As we approach the urban area, the scene is a chaotic mix of freely roaming goats and chickens, school children playing amidst mud and rubble, and idle buses where drivers gather for their morning coffee and catch up on the latest news. It truly feels like the last stop on the bus route out of Kathmandu.

In preparation for the challenging climb ahead, we're offered strong coffee and lentil stew for breakfast. I pass on the stew, a decision I soon regret as the walk demands a much-needed energy boost.

Finally, we begin our ascent. Oh my goodness, the path ahead is incredibly steep, almost vertical, and we've only covered four hundred metres so far. I try my best to hide my nervousness as I adjust my rucksack for a more comfortable fit. Seeing people walking downhill already makes me envious, and it's only been fifteen minutes into the journey. I'm already struggling to catch my breath, desperately trying to keep up with Sam and our expert guides, Norbu and Tshering.

Tshering, our lead guide, is a remarkable individual. At the young age of twenty-two, he has summited Everest not once, but twice! Tshering has experienced its unforgiving environment first-hand, including navigating treacherous crevasses in the Khumbu Icefall. We are incredibly fortunate to have him leading our trek, alongside his cousin Norbu, who hails from one of the Himalayan villages.

We pause at a waterfall atop the initial incline, providing a moment to catch our breath. As I look down at the cascading torrents of brown water crashing against the rocks, I spot faded red and yellow prayer flags fluttering in the wind. These colourful rectangular cloths, which we'll encounter numerous times along the foothills, hold blessings for the surrounding countryside and more.

Just as I reach for my water bottle, I notice the national park sign outlining our route through the wildlife reserve. It serves as a wake-up call, highlighting the trail's difficulty – something that wasn't mentioned in the guidebook's description of it as an 'easy introduction for beginners'. It's dawning on me that the remainder of this trek will be physically gruelling. On the upside, the trek promises a perfect introduction to the nature and culture of the Himalayan countryside, offering majestic views of mountains like Everest, Langtang, and Gaurishankar. I can only hope those views will be worth it as we press on.

Continuing along the route, we face rugged and irregular stone steps that wind their way up the mountain. My calves and knees are already protesting, and the high altitude leaves me increasingly short of breath as we painstakingly ascend, step by step. Suddenly, school children dart past us, effortlessly running downhill in nothing but flip flops, while families carry on with their daily activities. It's a humbling moment when an elderly couple overtake me uphill.

We stop for lunch in a village perched on the mountainside, and I glance back to see the mist rising over the foothills of Kathmandu, revealing the expanse of cultivated green terraces wrapping around the mountains. As we enjoy our meal, goats are herded up a muddy track masquerading as a road. It's at this point that I have my first encounter with leeches. I flick the little black creature off my skin, watching it leave a trail of blood as it scurries away. Now I understand why Norbu salts his trainers to deter these bothersome critters – they hate it!

While Sam seems to be tackling the trek with ease, eager to make more progress uphill, I'm feeling sick and would rather linger in the village a bit longer. After devouring a bag of crisps for a fleeting energy boost and succumbing to the pressure from others, we resume our upward journey. Sam has the advantage of having previously walked the more arduous Annapurna Circuit in the Himalayas, so he knew what to expect and was better prepared.

Step by step, I slowly haul myself up the mountain, always finding myself as the last person at each meeting point. What's most frustrating is that just as I catch up to others resting for a mere five minutes, they immediately set off again. At this point, I feel like I could crawl up the mountain on my knees. Norbu repeatedly assures me that our resting point is not far, even though it always seems further than he suggests.

Suddenly, through a clearing, I spot a large yellow sign displaying 2,195 metres (7,201 feet). To put it in perspective, Ben Nevis, the highest mountain in the UK, stands at 1,345 metres (4,413 feet). We've nearly doubled that height by the end of day one. Excitedly, we gather for a group photograph.

The terrain levels off, and we begin a brief descent towards Chisapani, the village where we'll be spending the night. The dense woods give way to a wide grassy plain, and the village reveals itself before us. Oh my goodness, it appears to be teetering on the edge of the mountain. The remnants of buildings damaged by the 2015 earthquake lean precariously, seemingly ready to tumble at any moment. The entire village appears to be slipping down the mountainside, and the tea house where we'll be staying perches perilously on the edge. Yet, I'm so exhausted that I couldn't care less – I could sleep anywhere.

My assigned room in the tea house offers little solace. It lacks lights, is damp, and carries the musty odour of mould. A constantly dripping tap fills a plastic bucket that requires regular emptying. The owners inform me that I'm lucky to have running water and a toilet, but the stench of ammonia sewage makes my stomach churn every time I step into the bathroom. The bed feels damp, but with two duvets, I manage to find some semblance of warmth. Thankfully, I discover an electric socket that works, allowing me to charge my mobile phone overnight. Despite the less-than-ideal conditions, I have no choice but to make the best of it.

Tshering suggested wearing flip flops in the evenings to let our feet breathe, but the biting cold makes me discard them in favour of my trusty walking boots. Aching, exhausted, and shivering, I find it difficult to sleep, plagued by nightmares of tumbling off the mountain.

28 May 2018

I wake up feeling freezing cold and hastily throw on the only jumper I have with me. I make my way down to the reception area for breakfast, where our Nepalese host warmly greets us with Tibetan bread, accompanied by strawberry jam and a pot of black tea. The bread has a slight sweetness to it and is perfect for dipping in the tea, providing a comforting start to the day before we embark on our journey once again.

To my relief, the inclines are not as steep today, and there are fewer daunting stone steps to navigate. As I glance back at the village from a distance, the precarious buildings appear even more perilous, making me realise how fortunate we are to have made it out safely. As we pass through oak forests and grassy clearings, with stunning views of the Langtang and Gaurishankar mountain ranges, this leg of the trek feels gentler. However, as is often the case with downhill walks, there is always a climb that follows. It's tough, especially with a fully loaded backpack. I had already lightened the load by parting ways with my flip flops in Chisapani, knowing there will be more items left behind as we continue across the Himalayas.

Two young stray dogs join us for a portion of the journey as we traverse a forested area en route to Nagarkot View Tower, situated at an elevation of 2,175 metres, offering expansive vistas. The dogs seem playful and enjoy our company, though their true intentions may be to safeguard their territory from intruders like us. Upon reaching the Tower, we hope for a panoramic view stretching from Annapurna in the west to Everest in the east, but our hopes are dashed by low-lying clouds that obstruct visibility. Even at our hotel, strategically positioned to capitalise on the magnificent views, we are greeted with an unfortunate lack of visibility. Dishevelled and exhausted after a long day's trek, the receptionist's offer of a cold beer is warmly welcomed.

I retreat to a room that, thankfully, is a significant improvement from the previous night's experience. I stretch out on the bed, allowing my weary limbs to rest, while keeping the windows open to let the refreshing mountain breeze wash over me. Dinner is served on the terrace, and despite the quick chill that descends in the evening, we decide to tough it out. A glass or two of wine later, I muster the courage to ask Norbu and Tshering, "Do you think I have what it takes to summit Everest?" Their response is immediate and filled with hysterical laughter, leaving no room for doubt – it's a resounding no!

Throughout the journey, I have been carrying my backpack up and down hills, but today is different. I finally ask Norbu for assistance. Earlier, I had a moment of frustration when I realised I couldn't bear the weight of the bag on my weak shoulders and low energy levels while conquering yet another hill. It's a valuable lesson learnt – never overpack a backpack. At Nagarkot, I part ways with more unnecessary items, including spare trousers, moisturiser, and extra socks. Even with the weight significantly reduced, it doesn't take long before Norbu is shouldering my bag once again.

29 May 2018

At the break of dawn, we set off on our trek to Dhulikel. Leaving Nagarkot behind, we pass by a vibrant display of locally crafted felt slippers neatly arranged, likely intended for sale to tourists on bus trips. Along the way, we tackle three modest climbs. By the time we approach the final hill, I'm on the verge of giving up and contemplate seeking a motorcycle ride from the nearest village. However, Tshering persuades me to persevere and even goes out of his way to find a slightly easier but longer route to the summit. I can sense their weariness with my complaints, as their eyes subtly roll upward, but they are relieved when I'm in a better mood while descending. Thankfully, I don't struggle with knee issues, so going downhill is a breeze for me – I feel like a nimble mountain goat!

During our descent, we are surrounded by hills adorned with prayer flags, adding vibrant pops of colour to the landscape. We pass by cheerful women working in the fields, their laughter echoing as they balance water carriers on their heads. Before reaching our destination, we encounter two women foraging, each carrying a sharp-bladed knife on their waistbands. They know a little English and pause to greet us before disappearing back into the bushes to resume their task.

As the path winds its way down to Dhulikhel, the scenery undergoes a transformation. It's disheartening to witness pathways littered with debris and discarded plastic rubbish. Upon entering the bustling town, we walk alongside a congested and polluted road, being cautious not to step onto it until we reach the hotel tucked away around a corner. My legs are stiff and unyielding, making it a challenge to rise from a seated position. Exhausted and sunburnt, I relish the taste of a cold beer, savouring the moment as we toast the end of our trek. With bottles clinking, I can't help but feel an invincible spirit within me, whispering that nothing can stop me now – Everest is the next conquest!

30 May 2018

Unfortunately, I have a flight to catch in Kathmandu, so I don't have the opportunity to explore Dhulikhel, which I've heard is a charming medieval town with many religious sites of interest. Yesterday's trek up the grimy and dusty arterial roads wasn't pleasant, and today's heavy rain adds to the misery during the two-hour drive back to the airport. Upon arrival, I'm swiftly ushered through the process as Sam eagerly prepares for more sightseeing. He's already frustrated with the delay, but for once I manage to check-in on time and there's no fog to disrupt the flight schedule. I'm heading back home with an invigorated spirit and feeling as fit as a mountain goat thanks to the high-altitude experience.

39. CINNAMON IN SRI LANKA

**London Heathrow Airport –
Bandaranaike International Airport, Sri Lanka**

Flight time: Ten hours and twenty five minutes

6 July 2018

It's great to be back in Sri Lanka for my third visit, and I can already see how much has changed since my previous trips. The road from the airport to Colombo, which used to be a slow journey through villages, has been transformed into a toll road, thanks to the Chinese investment and development projects. This new road has significantly reduced travel time and improved connectivity. Speaking of Chinese investment, they have been one of the major contributors to Sri Lanka's economy, fuelling growth and development in various sectors.

Even Colombo, the capital city, has undergone subtle changes. One noticeable difference is the absence of wild dogs roaming the streets, a welcome improvement for the city's residents and visitors.

Although my previous visits were more memorable for the extracurricular activities, such as partying on golden beaches and witnessing lively debates amongst colleagues in restaurants, I also had the chance to explore the natural beauty of Sri Lanka. One particular highlight was my visit to Kandy, nestled in the mountains. There, I embarked on an exhilarating train journey that remains one of the most memorable experiences of my life. The scenic views from the train windows were incredible, with waterfalls, cloud-covered mountain peaks, and lush tea gardens capturing my attention at every turn. The train ride defied conventional notions of safety, as people held onto handrails on open doors, braving deep gorges and numerous tunnels. As we descended towards Colombo, the glistening southern coastline came into view, adding to the awe-inspiring journey.

Being back in Sri Lanka brings back fond memories, and I'm looking forward to my time at the conference and exploring more of this beautiful country.

7 July 2018

The conference I attended today was rather unexciting, as it covered familiar ground on the topic of good government financial management. While there were a few more women in attendance compared to previous events, most participants were still men. The dry and technical subject matter failed to ignite any real enthusiasm, so I was relieved when it finally concluded.

However, as a token of appreciation for my contributions, I received a unique and thoughtful gift: ground and whole cinnamon. Sri Lanka is renowned for producing the finest cinnamon in the world, so this gesture holds special significance. It's a departure from the typical store-bought cinnamon, which tends to be cheaper and less refined. When I take a whiff of the cinnamon, I'm greeted by its earthy and aromatic scent. It's a fitting and pleasant way to conclude my visit to Sri Lanka, knowing that I'll have a tangible reminder of the country's rich cinnamon heritage.

40. ONE LAST TIME IN BEIJING

London Heathrow Airport –
Helsinki-Vantaa International Airport, Helsinki, Finland –
Beijing, China

Flight time: Nine hours and forty minutes

20 August 2018

Once again, I find myself at Helsinki-Vantaa Airport, waiting for my connecting flight to Beijing. Over time, I've become a frequent traveller to China, so I made the wise decision to invest in a multiple-entry visa that saves me from having to renew it for each individual visit. Additionally, I acquired a second passport, allowing me to travel on one while the other is at an embassy for visa processing. These strategies have proven to be quite convenient.

I appreciate Helsinki-Vantaa Airport for its relatively small size and ease of making onward flight connections. There are no long, sprawling corridors to rush through or endless security checks to endure. The airport is filled with charming shops selling locally made goods, providing a pleasant way to pass the time during layovers. It is a sharp contrast to Beijing Airport, which is of colossal proportions. The shuttle ride from terminal to terminal seems to go on forever, passing an endless row of tall red pillars until reaching terminal one passport control.

During my stay in Beijing, I will be located near the bustling business district, an area I have become quite familiar with over the years. My last visit was in 2017, where I had a series of meetings with universities and various government bodies. It was a whirlwind tour, so intense that when the border control officer at Heathrow scanned my passport, he looked at the destinations and miles covered in disbelief. We shared a smile and a high five. The itinerary of that trip went something like this: on day one I had meetings in Beijing; on day two I started at 4am and took a two-and-a-half-hour flight to Nanjing for a meeting with the university, returning to Beijing by 10pm; and on day three I flew back to the UK. It was truly a whirlwind experience.

Several memories stand out from my previous visit. Firstly, the heavy pollution in Nanjing was quite striking, obscuring the sun and affecting the overall atmosphere. Secondly, I remember enduring a long lunch where I had to engage in small talk while struggling to enjoy Chinese cuisine, which is not to my taste. The meal seemed never-ending, and my reluctance to consume much of the food on my plate attracted attention from both the chef and my colleagues. Long business lunches are one of my least favourite things, and I try to avoid them whenever possible. Lastly, I recall visiting the Museum of Audit, a rather peculiar experience. Yes, there is indeed such a museum, perhaps the only one of its kind in the world, and somehow I stumbled upon it. As we progressed through the museum's galleries, showcasing ancient, modern, and contemporary audit practices, I felt a migraine coming on. It was a revelation to learn that auditing has a history dating back three thousand years. With Chinese auditing stamps in hand, I boarded the flight back to Beijing.

Despite the challenges and peculiarities, each visit to China has its unique moments and I look forward to discovering what this trip has in store.

21 August 2018

I find myself occupied this morning with meetings with government officials from the Ministry of Finance in Beijing. The taxi drops me off in the leafy Xicheng District, a suburban area within the city centre. I wait on the corner of a junction until the government officials join me, and together we enter the building through security gates. As two additional officials join us, we commence our meeting to explore areas of mutual interest and potential cooperation. The Ministry of Finance in China is a massive institution, dwarfing the ministries of many other countries in terms of size and structure. As the meeting progresses, I struggle to fully comprehend the various functions and responsibilities of the different departments housed within the Ministry, some of which are economic or regulatory in nature, while others serve different purposes.

At one point during the meeting, they express interest in using a financial management assessment tool that we have developed to gauge their efficiency. I have the authority to offer them access to the tool and to work in partnership with them on its application. However, a caveat is presented – they desire a partnership but are unwilling to allow any independent assessment. They are not prepared to share the results with us. In hindsight, I should have expected such a stance from an authoritarian one-party state that values secrecy. I depart the meeting with the understanding that we will continue to engage in discussions in the spirit of mutual cooperation going forward. While the outcome may have been relatively fruitless, we conclude the meeting with the obligatory photo on the steps of the Ministry of Finance.

I flag down a taxi to head to my next meeting with the National Audit Office of China, which has become a familiar place for me during my visits to Beijing. This time, we are headed to the Xicheng District. The Audit Office serves as an important supervisory body outlined in the constitution and is an integral part of China's supervisory system for both the Party and the State. It plays an active role in upholding financial disciplines and combating corruption to foster clean governance.

As on previous occasions, the driver drops me off at the wrong entrance to the building due to the language barrier. Consequently, I find myself taking a long trek along the perimeter fence of a vast modern building in an industrial estate. The director of international cooperation comes to greet me, and we proceed to a spacious meeting room where I share international practices with several directors from across the audit institution. The topic that piques their interest the most is anti-fraud and corruption measures. Transparency International ranks China 66th out of 180 countries in its corruption perceptions index, so it is not surprising that they are keen to discuss this subject and remain open to hearing new ideas on how to address it. We conclude the meeting on a positive note, and I flag down a taxi on the ramp to head back to the hotel, looking forward to some late-afternoon time off-the-clock.

The driver and a guide arrive on time, and we embark on a visit to see the giant pandas at Beijing Zoo and explore the Summer Palace with its expansive gardens and lakes. These are two tourist attractions that I have yet to experience. As we arrive at the zoo, it is bustling with visitors who, like us, are eager to catch a glimpse of the highly esteemed giant pandas. These cuddly creatures, now endangered with an estimated global population of 1,200 and only 10% in zoos, captivate me. I find it difficult to put my camera away as I observe their incredible curiosity and playfulness. Some are sleeping while others engage in playful activities, especially the younger ones. I can't help but squeal with delight as I witness one descending a children's slide into a playpen.

Prior to visiting the zoo, I had initial concerns about the conditions in which the pandas are kept, but it seems that efforts have been made to accommodate their needs. Nonetheless, I would prefer to see them in their natural habitat, roaming freely. Jenny, my guide, seems impatient to move on to our next destination, the Summer Palace, built by Emperor Guangxu.

Jenny is a multilingual guide who fits tours like these into her studies. She shares that she is single, twenty-seven years old, and hails from the Huabei province in northern China. Having recently relocated to Beijing, she discusses the high cost of renting apartments in the capital and the tremendous social pressure imposed by her family to find a partner and marry. Her stories remind me of a book I read, *Leftover in China*, written by Roseann Lake, which explores women's roles in Chinese society.[19] The book even mentions instances where young women would hire men to pose as their boyfriends when visiting their parents. The higher the fee, the more convincing the charade, including holding hands in front of the parents.

Four decades ago, marriage was the norm in China and often the sole means of securing a livelihood for women. However, due to the one-child policy implemented for many years, the first generation of urban-born daughters were encouraged to pursue education and achieve success as if they were sons. Now, a significant majority of these women have chosen to postpone marriage or not marry at all. They struggle to find suitable partners and bear the label of 'leftovers', as described by Roseann. Jenny places herself within this category.

We enter the Summer Palace through grand wrought-iron gates on the eastern side. It is a gloriously sunny afternoon, and it seems everyone had the same idea as us – to stroll around the lake and wander through the Imperial Gardens. Jenny continues to share stories about life in China as we explore the vast gardens, exemplifying Chinese garden design principles. She explains that the Palace was destroyed during the 1860 war but was subsequently restored. Against the backdrop of hills and open water, we pass by a multitude of pavilions, halls, temples, and bridges, each possessing remarkable aesthetic value that compels me to capture their beauty through my camera lens.

---

19   Roseann Lake, *Leftover in China*, 2018

Jenny effortlessly rattles off the names of the buildings: the Tower of the Fragrance of the Buddha, the Tower of the Revolving Archive, Wu Fang Pavilion, the Baoyun Bronze Pavilion, and the list goes on. We decide to board a dragon boat to cross Kunming Lake, seeking respite from the heat and a chance to rest our legs. As we glide across the serene lake, we discuss topics such as affordable fashion and the prevalence of fake designer handbags and scarves in local markets. Jenny reveals that the best products are found behind closed doors and offers to take me to a trusted trader. I had made a pact with myself not to be tempted by cheap fashion, but I find myself intrigued by what lies behind those closed doors.

We arrive at an apartment block somewhere in the city and are greeted by a young man who appears to be in his early thirties. Jenny introduces us, and we take the elevator up to the 25th floor, entering an apartment filled to the brim with copies of the latest trends and designer handbags. Yves Saint-Laurent, Gucci, Burberry, Coach, DKNY, Longchamp, Mulberry – you name the brand, and there's a copy available. I feel a sense of nervousness, well aware that I am taking a significant risk by being here. If caught, I could end up in a Chinese prison due to my own choices. Against my better judgment, I decide to purchase a small, high-quality replica of a well-known branded handbag. There is no need for cash, as the trader conveniently has a card reader. With the transaction completed, I leave the building, clutching the handbag wrapped in a black bin liner. As we make our way out, we spot policemen patrolling the parking lot, further amplifying my anxiety as I hope to find our driver swiftly. Thankfully, they are traffic police, and we breathe a sigh of relief once we are back in the car.

22 August 2018

I wake up early to embark on another busy day, delivering a full-day seminar for accounting students at an accountancy school located on the outskirts of Beijing. Beforehand, I decide to go for a jog through the streets of the city and stumble upon a grand canal that cuts through the urban landscape. The canal walkways are pristine and adorned with newly planted trees, providing a picturesque setting for my morning jog. The rising sun casts reflections of the glass buildings on the water, creating a shimmering effect. The tranquillity of the surroundings prompts me to pause and sit for a while, but my peaceful moment is interrupted by a call from Aidan, who wants to check on how I've been.

The line of trees along the canal provides a refreshing shade, and I can't help but notice the positive changes in the city's environmental conditions. Over the years, Beijing, in particular, has faced criticism for high levels of air pollution. However, there is now a visible difference, with the city being greener and experiencing reduced levels of air pollution. This improvement is partly attributed to extensive afforestation efforts, with the forest coverage increasing from 13% to 44% since 1980 following decades of investment as reported by the Capital Greening Office.

Although I have visited this business and accountancy school before for meetings with the chancellor and a tour of the lecture theatres, one peculiar room caught my attention. Inside, I found a collection of TV screens and cameras, with Party observers monitoring each lecture room. My guide swiftly moved me along, and I didn't dare enquire about their presence, assuming they were there to ensure no subversion was taking place during lectures.

The seminar is being simultaneously translated, requiring additional concentration on my part, but it seems to be going well. The students are engaged and actively participate by asking numerous questions. I particularly appreciate how they organise themselves into groups and designate individuals within each group to answer questions. This approach facilitates smoother group discussions, as in my experience it can sometimes be challenging to get everyone to agree on a leader for feedback, questions, and note-taking.

After a brief lunch break, I am taken to a small room to rest and nap before the afternoon session. It feels somewhat peculiar being alone in a room, staring at the four walls for an extended period of time. The afternoon seminar is more enjoyable as the students become more confident in asking questions and actively contributing to the discussion. We conclude the session with lots of fun photos before I am transported back to the hotel to pack for my trip to Vietnam.

Arriving at the airport, I experience a slight concern regarding the wad of fresh GMB notes I'm carrying, along with the replica designer handbag. I worry about potential questioning during security control. Fortunately, I manage to pass through security without any issues, and a sense of relief washes over me.

41. FULL CIRCLE TO HANOI

**Beijing Capital International Airport, China –
Nội Bài International Airport, Hanoi, Vietnam**

Flight time: Three hours and forty minutes

23 August 2018

After a three-hour flight from Beijing, I find myself back in Hanoi reunited with old friends. However, my day begins with a setback as the hotel rooms are not yet prepared, leaving me and a group of tired holidaymakers stranded in the foyer. The group, recently arrived from the UK, looks dishevelled from their long-haul flight, but they are excited to embark on their dream vacation whilst I am here for work.

Following the delayed room preparations, I quickly move on to my first meeting with government officials, which is unsurprisingly dominated by men. Nevertheless, I hold my ground and contribute to the discussions. This is followed by a roundtable discussion where Asian accountancy professionals exchange views and practices. As an emerging economy, Vietnam has received significant donor funding for its development, and our focus is to share budget practices and financial reporting developments with government officials. To my delight, there is another female attendee in this meeting, marking a 100% increase in female representation. It makes me wonder how the conversation might change with more gender balance, possibly incorporating discussions on gender budgeting and its relevance to Vietnam.

After a long and intense afternoon of technical discussions, I manage to squeeze in a visit to the Women's Museum, which I had previously explored. The museum has undergone changes, becoming more commercialised with a new exhibition space and a visible entrance from the street. The current exhibition showcases costume design throughout the ages, displaying beautifully embroidered and colourful costumes. As always, I am tempted by the shop and end up purchasing another handmade scarf, which I cherish.

In the evening, I hail a taxi to meet my colleagues in the old part of Hanoi. However, I struggle to find the hotel bar they are in, and my phone battery is running low. Thankfully, a helpful shop assistant comes to my aid, and I locate my colleagues just fifty metres away. As we gather in the warm and sultry evening, I find myself tuning out of their shop talk, enjoying a glass of wine, and reflecting on my decision to leave behind my jet-setting lifestyle. After more than a decade of travelling the world, promoting public financial management, I feel it is time to explore new avenues and areas of interest, a secret I keep for now.

The three of us decide to move to a small traditional restaurant, where we share delicious small plates of food and indulge in copious amounts of wine. As the evening ends, our adventurous spirit kicks in, leading Sam and me to consider riding pillion on motorcycles back to the hotel. However, Rob intervenes and arranges for a taxi instead, ensuring our safe return.

24 August 2018

After a long night of indulgence, we start the day with more meetings before finding some time to explore the bustling streets of Hanoi's old quarter. The old quarter, a congested square filled with shops, is a popular destination for both business and tourists. However, the narrow and densely populated streets make it challenging for me to navigate and get my bearings. Shopkeepers try to entice me into their stores at every turn, and I have to skilfully manoeuvre through the traffic and crowds of people.

Feeling overwhelmed by the chaos, I decide to seek out my colleagues and find a place to eat. We stumble upon a lively outdoor restaurant that is packed with customers. The aroma of fresh herbs and noodles cooked with pork fills the air, instantly enticing us. We load up our plates with delicious food: fresh spring rolls, noodles with grilled pork, an assortment of herbs and chilies, and rice-batter pancakes. The flavours are exquisite and the Vietnamese herbs, such as mint, sweet basil, and dill, add an extra fragrant dimension to the salads that is hard to find elsewhere.

The meal is a delight for the senses, and the fast-paced service and engaging conversation make for a truly enjoyable end to the day. The combination of tasty food, vibrant atmosphere, and great company brings a sense of relaxation and contentment.

25 August 2018

Today is a day of freedom, where I can do as I please, and my destination of choice is Ha Long Bay. This magnificent bay is adorned with thousands of limestone islands and islets, emerging from the aqua green sea, their summits draped in lush rainforests. To kickstart my adventure, I hop on a motor scooter and become a passenger, zooming through the chaotic traffic of Hanoi. The thrill of it all is truly exhilarating, although I must admit, I have some concerns about insurance despite wearing a crash helmet.

After the scooter ride, I switch over to a small minibus, guided by an authoritative tour guide who seems more intent on disturbing tourists than welcoming them. Even before our journey commences, she already finds herself in a dispute with a couple of passengers, resulting in their departure from the bus. Customer service clearly isn't her strong suit.

Following a drive of over three hours, covering 125 kilometres, we finally arrive at the bay, greeted by a amazing sight of floating islands resembling fluffy meringues, stretching as far as the eye can see. We transition onto an old diesel boat, embarking on a journey to explore these mystical islands. The ever-changing scenery captivates me, making it impossible to put my camera away. We indulge in a delectable lunch below deck, feasting on langoustine and grilled fresh fish accompanied by a refreshing cold beer. Our first stop is the Trinh Nu Cave, also known as the Virgin Cave, named after a poignant love story of a young girl from a fishing village who found refuge on this island to escape an unwanted arranged marriage, as her heart belonged to another.

As I join the queue of fellow tourists, we venture through the heart of the cave, where glistening white stalactites gracefully descend to the cave floor. The coolness inside offers a welcome respite from the scorching sun outside. At the centre, a towering chasm reveals a beam of white sunlight seeping through a crack in the wall, illuminating the intricate shapes and sculptures adorning the cave walls. Upon emerging back into the sunlight, we pass by rows of budget souvenir shops as we continue our journey to the next destination.

Our next stop takes us to Cua Van, one of the old floating fishing villages in the bay. We disembark onto a pontoon and board a traditional fishing boat, gracefully gliding through the small islands and hidden inlets. This tranquil village, surrounded by jade green waters and limestone formations, remains a hub of fishing activity. Colourful rafts and boats intermingle, creating a captivating scene. As we navigate through narrow water caves and encounter serene pools enveloped by lush greenery and rock formations, I find myself immersed in a state of complete relaxation, eagerly observing the birds and wildlife that grace this picturesque setting.

During our return journey, the once-standoffish guide has surprisingly transformed into a more friendly demeanour. She insists on assisting me in finding the best coffee in Vietnam, leading me to her friend's shop, where she undoubtedly receives a commission. Walking through Hanoi's old quarter, we pass street-food vendors displaying an array of mouth-watering delicacies, from oysters to crabs, laid out and ready to be cooked. I seize the opportunity to purchase some coffee beans before bidding farewell, hitching a spontaneous ride on the guide's motor scooter, much to her surprise.

And so, my travels, at least for now, come to a close, marked by a splendid day and my final business trip. Returning full circle to Hanoi, where my first journey began in 2009, I can't help but cherish the entire travel experience. It has been a joyous adventure, allowing me to grow as an individual and acquire invaluable knowledge along the way. However, my story doesn't end here; there are undoubtedly more chapters awaiting me in the future.

42. REFLECTIONS IN THE COVID YEARS

2020-2021

The COVID years brought about a significant shift in the world, and international travel came to a halt. Surprisingly, I didn't mind this change at all, considering the extensive travelling I had done over the years. Staying at home and avoiding planes and trains was a welcome respite. It provided me with an opportunity to reflect and indulge in my love for travel through the pages of countless books. During this time, I also began writing this journal and focused on establishing Public Finance by Women, an organisation aimed at supporting women in public finance and promoting gender equality. The empowerment of women in finance was a cause close to my heart, given my own experiences detailed in this journal. However, launching a new organisation during such uncertain and challenging times proved to be a daunting task.

Fortunately, the power of technology and social media allowed me to leverage the connections and networks I had built, enabling Public Finance by Women to achieve remarkable feats during the COVID years. We successfully launched the first international mentoring scheme, which not only kept me engaged and occupied but also addressed an important issue. The advancements in technology have transformed the way we work, making my previous roles look vastly different with reduced corporate travel, lower costs, and a minimal carbon footprint. It would have allowed for a more strategic approach to attending far-flung meetings and events.

However, while technology offers numerous advantages, it also has its limitations. The absence of personal, cultural, and social interactions that come with face-to-face meetings can hinder certain aspects of work. Building trust, developing strong business relationships, and finding common ground often require in-person interactions. Professional networks play a vital role, and sometimes only the connections forged during physical meetings can have a lasting impact.

As the memory of COVID gradually fades, I unexpectedly find myself embarking on a new tour to the Western Balkans, working with a supranational organisation. Although I had thought my days of business travel were behind me, this opportunity presents itself and I'm unsure how I will feel about it. It seems that my business travelling boots may not have been retired after all.

43. LOST LUGGAGE IN BOSNIA AND HERZEGOVINA

**London Gatwick Airport –
Vienna International Airport, Austria –
Sarajevo International Airport, Bosnia and Herzegovina**

Flight time: Five hours

19 September 2022

I find myself back at Gatwick Airport, and it seems like so much has changed since my last visit. Self-service check-ins have become the norm, but the frustratingly long queues at the baggage drop-off remain unchanged – airports are just not my favourite places. The walk to the departure gate feels like the longest ever, as it's located at the far end of the terminal building. Thankfully, the flight to Vienna is short, but with only a half-hour turnaround time at Vienna International Airport to catch my connecting flight, it's going to be a challenge. Unfortunately, the flight from London is delayed, adding to the stress.

Upon arriving at Sarajevo International Airport, I discover that my bags haven't made it. I stand by the conveyor belt, watching it go round and round, with only a single battered, orange suitcase making its way down.

I'm not alone in this predicament, as other passengers are also missing their luggage. The airport staff try to be helpful, but they don't explain that lost baggage is a common occurrence here and it's unlikely that I'll see my bags again. So here I am in Sarajevo without a change of clothes or my essential business items. I feel completely stranded.

At least I have a couple of days to regroup and explore Sarajevo before leading the training course on gender and public financial management. Sarajevo is a small capital city situated in a valley surrounded by mountains, with the Milyatska river running through it. The business centre of the city is the central point where people work and gather, while the residential districts, known as *mahalas,* cling to the hillsides. Each quarter represents a different culture and religion, with the Muslim, Jewish, Catholic, and Eastern Orthodox quarters peacefully coexisting side by side. After a quick orientation of the old town and business area, I decide to grab a Sarajevan kebab for sustenance before contacting the lost baggage office at the airport to check for any updates. Unfortunately, there is no positive news to be found.

20 September 2022

Wearing the same clothes for another day, I can't help but feel a bit grubby and frustrated. In the past twenty-four hours, I've already spent a significant amount of money buying essentials like underwear and toiletries. I've exhausted my options of wandering around the old town, sipping Bosnian coffee, and making repeated calls to the lost baggage office. Unfortunately, there is still no news about my missing bags, and the excuses I've been given range from staff absences in Vienna due to COVID to mass redundancies and the airport's inability to handle the daily influx of flights.

In an attempt to calm myself down, I decide to buy a book about the recent history of Bosnia and Herzegovina. It proves to be an interesting read, shedding light on the country's past as part of the former Yugoslavian federation under the leadership of Josip Broz Tito. After Tito's death in 1980, Yugoslavia dissolved, leading to the formation of six independent countries, including Slovenia, Croatia, Bosnia and Herzegovina, Montenegro, Serbia, and North Macedonia. However, unresolved tensions between ethnic minorities seeking independence within these new states fuelled the outbreak of the war in 1992. While most conflicts were resolved through peace accords and the international recognition of new states, the region suffered immense loss of life and severe economic damage. It's clear that some of these tensions still persist today, particularly between Kosovo and Serbia. Serbia does not recognise Kosovo as an independent state following Kosovo's declaration of independence from Serbia in 2008.

I take a moment to reflect on what I've learnt before reluctantly making yet another call to the lost baggage office, only to be met with the same disappointing response – still no news.

21 September 2022

This morning, I finally receive the long-awaited news that my bags have arrived at the airport and are ready for dispatch. However, as the day progresses, I become increasingly frustrated as the bags still fail to make their way to the hotel. It's now 5pm, and there's no sign of my belongings. Growing impatient, I decide to make another call to enquire about the delay, only to discover that the annual city marathon had obstructed the driver's access to the hotel. It seems that he was probably enjoying a leisurely Sunday lunch while I anxiously awaited my bags. Finally, at 8pm, my bags are delivered to the hotel, bringing an end to this tiresome ordeal.

22 September 2022
The journey to Neum on the Adriatic coast proves to be quite an adventure. I meet up with my colleagues, Jess and Emily, whom I've only interacted with through emails until now. Together, we embark on a four-hour drive through the scenic, mountainous countryside. The winding roads make for a bumpy ride, and we find ourselves tossed about in the car with every hairpin bend. Jess, full of enthusiasm, engages in a lengthy discussion about the UK royal family and shares her opinions on the succession to the throne. I kindly explain that I was hoping for a break from all the pomp and ceremony, prompting laughter from Jess as she changes the topic.

    As we approach Neum, the stunning blue waters of the Adriatic Sea come into view. Emily informs us that during the Yugoslav era, this coastal town served as a retreat for the communist elite. Some aspects of the infrastructure may be in need of an update, but overall, Neum is a charming and picturesque seaside resort. Despite the temptation to indulge in swimming and sunbathing, I remind myself that I am here for a training course that will span the next couple of days.

23 September 2022

The training event on gender and public financial management is off to a promising start today. I find myself settling into a rhythm of multitasking, seamlessly transitioning between the slides translated in Bosnian and my handheld English version.

During a coffee break, my mind drifts to a publication I once read titled 'How we survived communism and even laughed'.[20] The author humorously highlighted the absurdity of considering women as having influence in public discourse in Eastern Bloc countries like the former Yugoslavia. Communism was not conducive to open dialogue, let alone granting women a voice. This brings to mind the question of how much has truly changed for women in Bosnia and Herzegovina since the end of communism. The author argued that remnants of the communist era persist in people's behaviour, their expressions, and their ways of thinking. I contemplate how much of this observation still holds true in modern-day Bosnia and Herzegovina.

---

20 Slavenka Drakulić, *How We Survived Communism and Even Laughed*, 1992

24 September 2022

The day feels especially arduous, especially when I catch glimpses of the bright sunshine outside, reflecting on the glistening water as the last few tourists of the season leisurely soak up the sun by the pool. Nevertheless, the three groups of participants from various regions of Bosnia and Herzegovina remain engaged and committed to the training topic. To keep my energy levels up, I consume several cups of Bosnian coffee throughout the day as we approach the final stretch. Eventually, it's time to wrap up and prepare for the winding journey back to Sarajevo.

A couple of weeks later, an unexpected email lands in my inbox with the subject line: 'How are you fixed for Kosovo?'. And just like that, my journey continues.

44. THE PRISTINA BEARS OF KOSOVO

**Luton Airport –
Pristina International Airport, Kosovo**

Flight time: Two hours and fifty minutes

6 December 2022

Braving the cold snap in London, I make my way to Luton Airport. Due to train strikes and a late-evening return flight, I have no choice but to drive myself. As I approach Luton, a sense of dread washes over me, knowing that I will be stuck in an airport that closely resembles a bus shed. I keep a close eye on the flight board, anxiously awaiting instructions to proceed to the departure gate. Despite the imminent departure time, the information board instructs passengers to sit back and relax. If I follow those instructions, I'll surely miss my flight. Determined not to let that happen, I hustle my way to the gate.

Finally, I arrive in Pristina, the capital city of Kosovo, greeted by a dreary sky. According to the guidebook, Pristina is a city that can be explored in just one day, so I plan to make the most of my time here before diving into business matters.

7 December 2022

I hop into a taxi and make my way to the bear sanctuary located in the hills just outside of Pristina. Thanks to a legislation passed in 2010, it is now illegal to keep bears privately, leading to the establishment of this sanctuary a few years later. The sanctuary provides a safe haven for approximately twenty bears that were rescued from deplorable conditions, such as cramped cages next to restaurants, where they were exploited to attract customers. I'm relieved to know that these bears can now roam freely within a large, forested area, enjoying a life of safety and proper care.

Embarking on a 1.8km circular walk, I begin my journey to observe these magnificent creatures in their new habitat. The climb is quite challenging, but it's worth it when I spot the first bear – a majestic caramel-coloured giant. Even with a sturdy metal fence separating us, I'm cautious not to get too close, while the bear curiously watches me. Continuing my walk, I encounter another bear peacefully basking in the sun with her paws upturned. Her name is Hope, and she was rescued in 2013 after being cruelly taken away from her mother and confined to a three-metre cage. Witnessing her now living her best life, receiving the utmost care, is truly heart-warming.

However, the walk is not without its concerns. I can't help but notice numerous warning signs cautioning visitors about the horned viper snake and what to do in case of a bite. With each step, I find myself glancing down, double-checking for any signs of these elusive creatures basking in the sun.

Returning to Pristina, I immediately embark on a walkabout to explore the city. My first stop is the National

Library of Kosovo, a striking building that catches my attention. Designed by Croatian architect Andrija Mutnjaković, the library's unique appearance has sparked controversy. With its brutalist architecture featuring multiple domes and windows of different sizes covered in a metal fishing net, it has gained notoriety for being considered 'ugly'. Inside, the library boasts elegant marble floors and has the capacity to house around 2 million volumes of books. The library holds great historical significance for Kosovo, having provided shelter for refugees from Bosnia and Herzegovina and Croatia during the Yugoslav wars, amongst other events.

Continuing my exploration, I visit the Kosovo National Art Gallery, established in 1979. The current exhibition, titled 'ordinary idylls of modern life', showcases the vibrant and textured paintings of artist Jeton Gusia. From scenes depicting beekeeping to everyday work commutes, the artwork is captivating and can be appreciated in a short amount of time. A short walk from the gallery takes me to the new Cathedral of St. Mother Teresa, a modern cathedral adorned with vivid stained-glass windows. From there, I stroll along Mother Teresa Boulevard, a bustling pedestrian walkway that seems to be the heart of Pristina. I take a seat at a café, immersing myself in the local atmosphere and indulging in some good old people-watching as I savour my coffee.

As I enjoy my beverage, I can't help but reflect on certain aspects of the city that hark back to Eastern Europe before the collapse of communism. The presence of groups of waiters aimlessly lingering in restaurants, a smoking culture that remains prevalent, and a city centre dominated by five-story Khrushchev blocks – all reminiscent of a low-wage economy and Eastern Bloc influences. However, on the outskirts of Pristina, newer developments are emerging, giving the impression of a more Western European city.

To conclude the day, I decide to conduct a practise run to the venue where the training event will take place over the next two days.

8 December 2022

I arrive early for the training event, eagerly awaiting the arrival of participants from Albania, Kosovo, and North Macedonia. Their enthusiasm to learn more about how public financial management can contribute to gender equality is evident. Most of them speak English as a second language and are kind enough to ask their questions in English, minimising the need for simultaneous translation.

Our discussions about achieving gender equality resonate with similar conversations held in other countries. However, there are unique aspects specific to Kosovo, such as the struggle for legal recognition of war-time sexual-violence survivors. It is disheartening to learn that it took nearly two decades after the conflict between ethnic Albanians and Serbian forces for comprehensive efforts to be made by the government, civil society, and survivors to implement reparations programs for the thousands of women who endured conflict-related sexual violence. A significant milestone was reached in 2017 when the Government of Kosovo allocated a budget for the process of recognising and verifying the status of war-time sexual-violence survivors, granting them legal recognition and rights. Witnessing the progress being made in this regard is encouraging.

Our discussions also bring to mind the tongue-in-cheek point made by author Slavenka Drakulić, who humorously highlighted the absence of tampons as an example of how communism failed to meet the basic needs of half the population during its seventy-year existence.[21] Though light-hearted, it serves as a reminder of the challenges faced by women in Eastern Europe during that era.

After two busy days, I find myself exhausted with a pounding headache as I board the long-stay car park bus. However, my fatigue is further compounded when I discover that I am charged an exorbitant £100 to exit the car park. It's a frustrating end to the journey.

As I hit the road on the M1 at midnight, I'm grateful to be heading home, ready to rest and reflect on the impactful discussions and experiences of the past few days.

---

21 Slavenka Drakulić, *How We Survived Communism and Even Laughed*, 1992

45. TRAVELLING IN MY SHOES

Who knows where my next adventure will take me, but I have a feeling there will be more stories to tell and lessons to learn. I consider myself incredibly fortunate to have had the opportunity to travel and experience so much throughout my career. Travel has enriched my life, opening my mind to diverse viewpoints and different ways of doing things dictated by various cultures. If you ever get the chance to travel for work or pleasure, I highly recommend taking it!

Being a businesswoman on the move has shaped me as a person over the years. It has made me more open-minded, willing to explore alternative approaches and listen to a wide range of perspectives. My ability to truly listen has been honed through my travels. I have developed a deep respect for different cultures, even though I may have made a few missteps along the way or faced challenging situations. I now recognise more than ever the value of diversity and inclusion, understanding how these can foster problem-solving and the importance of equal representation when bringing ideas into a room filled with individuals from diverse backgrounds.

Through my global travels, I have gained a better understanding of the geopolitical landscape, both in developed and developing countries, and the unique challenges they face, from economic growth to sustainability. I have witnessed the power of bringing experts together to share ideas and experiences in order to address global and local issues. Furthermore, I have come to appreciate the significance of forming deep bonds and building high-level trusting relationships with people from different backgrounds and cultures in order to bring about meaningful change. Personally and professionally, this journey has transformed me.

There are still areas where I hope to see progress, particularly in achieving greater gender equality. It has been an interesting experience working in a predominantly male-dominated environment but it is the twenty-first century and there should be more women in leadership roles. My career has not been without its challenges, as I have described in this journal. I have faced obstacles not being 'one of the boys'. Overall, however, my experience has been that, throughout my travels, male colleagues have respected my skills and professional status and in most cases my gender has not proved to be an issue in what is still, at least to some extent, a male-dominated business world. It is crucial to have more women in decision-making positions, and I will continue to advocate for this and improvements in policymaking to ensure gender considerations are taken into account.

On a lighter note, I have learnt the importance of having fun and taking care of my well-being, especially when travelling as extensively as I have. I believe I struck a balance in most of my journeys. And it's perfectly fine to have a bit of self-deprecating humour!

Looking back, as a female business traveller, I took a few unnecessary risks, like planning flights that arrived in the middle of the night in potentially unsafe countries. I wouldn't repeat those risks now, perhaps because I have become more risk-averse with age – I can't say for certain. Nevertheless, I admit that I often experienced moments of anxiety.

And so, my days of extensive business travel are coming to an end. It has been a truly remarkable experience for me, arriving in countries with only a basic understanding of their customs and practices. The lessons I learnt from my experiences couldn't be found in any travel guide. Who would have thought that I would struggle to cross a road in Vietnam without putting my life at risk or feel the need for a bodyguard in Nairobi!

I have cherished every single moment of this incredible journey!

ACKNOWLEDGEMENTS

Writing this journal has been a significant challenge, and I am deeply grateful to Dr. Aidan Rose, my husband. His unwavering support during my extensive travels and his patience in handling my late-night frustrated phone calls when things didn't go as planned have been invaluable. I have also lost count of the number of drafts I asked him to read, almost pushing him to the edge.

I extend my heartfelt thanks to Ann Shore, John Davies, and Derek Elliott, who, unbelievably, are three of my former bosses. Their willingness to review drafts and provide valuable insights and constructive feedback is deeply appreciated. Our longstanding, professional relationships have blossomed into solid friendships, although I must admit that this journal may have tested their limits!

To Mitzi Wyman, you are a shining star for consistently motivating me to persevere with this journal, even during the most challenging times. I would like to express my gratitude to Barbara Grunewald, an intrepid gypset and a dear friend who always offers thoughtful feedback. Regardless of her location in the world, I know I can always rely on her support. John Beattie, thank you for reading and providing feedback on the very first draft, drawing from your experience as a publisher in your own right.

Sumita Shah, as always, you have been amazing, and I hope you appreciate your pseudonym in the journal. Juliet Munro, your critical eye may be tough, but I value and have considered all of your comments. Penny and Kate Steed, thank you for taking the time to read my work. Kate, your enjoyment of the journal, particularly as a young woman, brought me great joy.

To anyone I may have inadvertently omitted, please accept my sincere apologies. I raise a glass of shampanskoye in your honour, with gratitude for your contribution and support throughout this journey.

MY BUSINESS TRAVEL SCHEDULE 2009-2023

Country	Year				
Austria, Vienna	2011	2017			
Argentina, Buenos Aires	2011				
Bangladesh, Dhaka	2012	2016	2017		
Barbados, Bridgetown	2013				
Belize, Belize City	2010				
Belgium, Brussels	Three times a year				
Sarajevo, Bosnia and Herzegovina	2022				
Canada – Ottawa, Toronto	2013	2016			
China, Beijing, Xi an	2012	2013	2016	2017	2018
Cyprus, Nicosia	2010				
France, Paris	At least once every year				
Hong Kong	2016				
Hungary, Budapest	2016				
India, Kolkata, Hyderabad	2013	2016			

Country	Year				
Italy, Rome	2014				
Jamaica, Kingston	2011	2014			
Japan, Tokyo	2003	2015			
Kenya, Nairobi	2011	2016			
Kosovo, Pristina	2022				
Luxembourg	2011	2017			
Maldives, Malé	2012				
Malaysia, Kuala Lumper	2016				
Nepal, Kathmandu	2013	2015	2018		
Northern Ireland	2015				
Pakistan, Islamabad	2016	2018			
Philippines, Manila	2014	2017			
Russia, Moscow	2010				
South Korea, Seoul	2011				
Sri Lanka, Colombo	2012	2014	2018		
South Africa, Johannesburg	2013	2014			
Switzerland, Geneva	2015				
Thailand, Bangkok	2012				
Turkey, Istanbul	2012				
Trinidad and Tobago, Port of Spain	2011	2014			
US, Washington, DC	2014				
Vietnam, Hanoi	2009	2018			
Zambia, Livingstone	2010				
Zimbabwe, Harare	2018				